SOFTWARE NETWORKING

Volume 1, 2016

SOFTWARE NETWORKING

Editor-in-Chief: Kuinam J. Kim, Kyonggi University, South Korea

Associate Editor: Hyuncheol Kim, Namseoul University, South Korea

Editorial Board
Ana Paiva, University of Porto, Portugal
Andrea Ceccarelli, University of Florence, Italy
Buseung Cho, Korea Institute of Science and Technology Information (KISTI), South Korea
Byung Yun Lee, Electronics and Telecommunications Research Institute (ETRI), South Korea
Chunqiu Zeng, FIU, United States
Chun-Yi Su, Concordia University, Canada
Daniel Bo-Wei Chen, School of Information Technology, Monash University Sunway
Campus, Australia
Derek Doran, Wright State University, United States
Edelberto Franco Silva, Fluminense Federal University, Brazil
Javid Taheri, Karlstad University, Sweden
Johann M. Marquez-Barja, Trinity College Dublin, Ireland
Longzhi Yang, Northumbria University, United Kingdom
Mahasak Ketcham, King Mongkut's University of Technology North Bangkok, Thailand
Michel Mainguenaud, Institut National des Sciences Appliquees Rouen, France
Michele Risi, University of Salerno, Italy
Mu-Song Chen, DYU, Taiwan
Pascal Lorenz, University of Haute Alsace, France
Reza Malekian, University of Pretoria, South Africa
Rossitza Goleva, Technical University of Sofia, Bulgaria
Sandro Leuchter, Hochschule Mannheim University of Applied Sciences, Germany
Sharmistha Chatterjee, Florida Atlantic University, United States
Tae Yeon Kim, Electronics and Telecommunications Research Institute (ETRI), South Korea
Terje Jensen, Telenor, Norway
Tomas Cerny, Czech Technical University, Czech Republic
Wonhyuk Lee, Korea Institute of Science and Technology Information (KISTI), South Korea
Yong Xiao, University of Houston, United States
Yin-Fu Huang, National Yunlin University of Science and Technology, Taiwan
Zeyar Aung, Masdar Institute, United Arab Emirates

This journal is published in collaboration with the iCatse organization: www.icatse.org

Aims and Scope
Software defined networking (SDN) and network functions virtualization (NFV) technologies have received significant attention over the last few years. In contrast to traditional networks, SDN decouples the control and data planes through an open and standard interface to enhance programmability and flexibility of carrier grade networks.

Published, sold and distributed by:
River Publishers
Alsbjergvej 10
9260 Gistrup
Denmark

River Publishers
Lange Geer 44
2611 PW Delft
The Netherlands

Tel.: +45369953197
www.riverpublishers.com

ISSN 2445-9739 (Online Version)
ISBN 978-87-93519-52-7

SOFTWARE NETWORKING

Volume 1, 2016

A Study on Trustworthy Cyber-Physical ID/Location Mapping on IoT and NFV

Hyeontaek Oh[1], Sanghong Ahn[1], Jinhong Yang[2] and Jun Kyun Choi[1]

[1]School of Electrical Engineering, KAIST, Republic of Korea
[2]Information and Communications Department, Korea Advanced Institute
of Science and Technology, 291, Daehak-ro, Yuseong-gu, Daejeon 34141
E-Mail: {hyeontaek; ancom21c; sunupnet}@kaist.ac.kr; jkchoi@ee.kaist.ac.kr

Received 20 January 2016; Accepted 1 March 2016;
Published 15 March 2016

Abstract

Information and Communication Technology (ICT) based industries, which are growing with advance of ICT technologies, are operated under trustworthiness of its ICT infrastructure. The identity trust targets the whole entities within the Internet of Things (IoT) and Network Functions Virtualization (NFV) environment that can be the creator, broker, modifier, and the end user of the data. Considering the dramatic growth of IoT service industry, its scalability, robustness, and efficient management are worth to be studied. Originally, conventional IDs for cyberspace are designed to consider having a relationship with physical address information. However, because of the wide-expansion of cyberspace, they are now very loosely-coupled. In this paper, the necessity of trustworthy cyber-physical mapping and its architecture would be discussed. The usage and demand of Cyber-Physical ID/Location mapping (CPID) will be discussed based on two use cases of real service environment. Furthermore, the requirements of CPID are introduced. A new CPID is needed to ensure trust location and trust identification, which includes location information explicitly for public goods in IoT and NFV environment, to make it easy to be recognized and used. Currently, related issues are under discussion in various standard organizations. To develop standards for trusted ICT infrastructure, it is needed to develop global CPID system by discussing ID/Location issues in global cyber identification or geolocation standardization groups.

Journal of Software Networking, 1–18.
doi: 10.13052/jsn2445-9739.20161.001

Keywords: trustworthy, cyber-physical, identity, locator.

1 Introduction

Industries on e-commerce, social services and the Internet of Things (IoT) are dramatically increasing, security and trust issues of cyberspace get emerged in multiple perspectives. Information and Communication Technology (ICT) based industries, which are growing with advance of ICT technologies, are operated under trustworthiness of its ICT infrastructure. In Seoul Conference on Cyberspace [1], they emphasize the importance of creating trustworthy cyberspace environment. Also, in World Summit on the Information Society forum 2015 [2], one of the important issues is "Building trust in cyberspace." Moreover, ITU-T Study Group 13 has been launched "Correspondence Group on Trust" for discussing trust provisioning for ICT environment [3].

The Internet of Things (IoTs) aims to provide whole-connected services based on ICT technology. Therefore, it is not only important to ensure security among IoT entities but also overall trust of data-driven services. It has led many researches related with Trust IoT environment to be done. The major issues of IoT Trust can be categorized into data perception trust, data fusion and mining trust, data transmission and communication trust, and identity trust [4]. The data perception trust and the data fusion and mining trust are related to the interpretation and use of data. The data transmission and communication trust is about delivery and distribution. The identity trust targets the whole entities within the IoT environment that can be the creator, broker, modifier, and the end user of the data. Considering the dramatic growth of IoT service industry, its scalability, robustness, and efficient management are worth to be studied.

Current forms of IDs in IoT services are designed for a device level, or oriented from a certain service to provide methods of object classification and recognition. For example, the identifier of EPCglobal standards adopted long hash code based combination, which is not suitable for human [5]. Especially, when IoT devices are widely deployed for public civil service (e.g. smart city, smart transportation, or disaster relief systems); if IoT devices and services utilize hash based identifiers, the users feel difficulties to recognize available services. It means that users, who unfamiliar with related services, are hard to intuitively identify the information from devices and services; and they are not able to recognize services and applications in cyberspace using physical addresses, which are one of the most popular identification mechanisms for human. In addition, even the virtualized network resources (e.g., routing, firewalling, load balancing, WAN optimization, etc.) need trusted ID system so that can be provided and utilized on the NFV environment. To build trusted connectivity between physical resources and virtualized ones, well designed cyber-physical ID structure is required.

Originally, conventional IDs for cyberspace are designed to consider having a relationship with physical address information. However, because of the wide-expansion of cyberspace, they are now very loosely-coupled [6]. As the number of networked devices and sensors is dramatically increased, and designed mapping between cyber ID and location is not working anymore.

Current loosely-coupled ID systems for cyberspace are not suitable for "cyber to physical" and "physical to cyber" based entity discovery and interoperating approaches which are required in cyber physical system (CPS) services [7]. These kinds of ID systems are not intuitive for users to identify a service that is related to devices/things in specific location, also it is hard to determine that resource of cyberspace affects actual location of physical area. Since everything has a location, therefore location information needs to be carefully described. In case of disaster or usage of public civil service, the infrastructure needs to provide methods that users could recognize and use resources, services and devices in the physical service area. For this purpose, a trustworthy cyber-physical identification and recognition system is required.

In this paper, the necessity of trustworthy cyber-physical mapping and its architecture would be discussed. To discuss these issues, current works for cyber-physical recognition and interoperation would be reviewed based on conventional cyber ID system. The necessity of cyber-physical mapping would be discussed with various scenarios in IoT and NFV environment. Also, current issues on cyber-physical ID/location mapping would be reviewed, which are discussed in ITU-T Trust Community Group. Based on them, the direction of studies and standardization for cyber-physical ID/location mapping and governance would be suggested.

2 Related Work for Cyber-Physical Identity and Location Mapping

Early ICT infrastructure was designed for a relatively small network of computers, where mobility was rare and participating nodes relatively homogenous. As a consequence, an early design decision established cyber ID as both a means to identify the end-point of communication and to specify its location within the network. However, when a number of cyber ID were dramatically increased, which caused by widespread of Internet and other ICT infrastructures, the implicit specification of this dual role for cyber ID has unforeseen at the time consequences.

In this section, we review history of physical address system and cyber ID system and their relationship, and also we review current issues of IoT environment based on cyber-physical ID/Location mapping perspective.

2.1 Physical Location Addressing Structure

Physical address system is widely used in everywhere, and the format of physical address is different based on their region. Until the advent of modern postal systems, most houses and buildings had no specific address, however, it has been commonly used after postal systems are introduced. Regarding physical address, in 1974, first version of ISO 3166 standard was published by the International Organization for Standardization (ISO) that defines codes for the names of countries, dependent territories, special areas of geographical interest, and their principal subdivisions [8]. Also, ISO standardized the universal postal union letter post regulations for unifying format of physical address. However, some countries kept their old address layouts and a few years ago the entire harmonization work collapsed and the standard specifications were declared deprecated. A postal code also is widely used, but every countries use in different way (in number 3-digit to 10-digit, in alphanumeric 6-digit to 8-digit). Today, there is no global unified standard for physical address system, each countries use their own format.

Satellite system is now utilized to geography survey, therefore, global positioning system (GPS) based location-based services are widely used. To standardize geo-location information, ISO and open geospatial consortium (OGC) has been developed related standards about protocols and interfaces for utilizing geospatial information in cyberspace domain.

2.2 Cyberspace Identification Structure

2.2.1 Telephony

A numbering plan for worldwide public switched telephone network is defined in ITU-T E.164 standard [9]. Plan-conforming numbers are limited to a maximum of 15 digits. In telephone number plan, each countries has their own country calling code and geographic area codes which can briefly describe the location information. However, a number of non-geographic area codes are increasing because of development of mobile telephony and

Internet telephony. It makes weaker relationship between telephone number and location information.

2.2.2 Internet

IP addresses (IPv4 and IPv6) are widely used for indicating hosts in the Internet. Since a number of connected devices are small, the IP can have both a means to identify the end-point of communication and to specify its location within the Internet.

However, as a number of connected devices in the Internet has been increased, this relationship has been changed. For example, a certain area has multiple classful IP addresses, and the number of private network is increased by network address translation (NAT) technique. These kinds of changes make hard to represent both cyber ID and physical location for IP address. Moreover, DHCP (Dynamic Host Configuration Protocol) based IP allocation makes it hard because IP address of network host periodically changed.

IPv6, which is developed to overcome IPv4 address exhaustion, is defined in IETF RFC 2460 [10]. IPv6 (and 6LoWPAN) uses stateless address auto-configuration scheme (which is similar to DHCP in IPv4), so network prefix is not able to bind with location information [11]. Domain name system (DNS) translates domain names (mostly using URL format), which can be easily memorized by humans, to the numerical IP addresses vice versa. Domain name space is consists of a tree of domain names, and the tree sub-divides into zones which is beginning at the root zone. DNS can represent some geolocation information by using DNS zones (e.g. country domain zones line ".us", ".eu", etc.) and domain name. However, since a number of domain names with general DNS zones (like ".com", ".net") are dramatically increased, domain name becomes loosely-coupled with its geolocation information.

2.3 Identification Methods for IoT Environment

The IoT environments aim at providing new services by connecting and collaborating various physical things in cyberspace domain using ICT infrastructure. For this purpose, the methods for identifying and discovering existing services and devices are needed.

For identification systems, EPCglobal (Electronic Product Code global) standards provide EPC based ID, which is used in RFID (Radio Frequency Identification) environment. In the Web of Things (WoT) concept in World Wide Web Consortium (W3C), WoT concentrates on utilizing resources and information that are able to access via HTTP (HyperText Transfer Protocol) only [12]. To identify these Web resources, URI (Uniform Resource Identifier) is used [13]. In CPS area, cyber-physical ID/Locator mapping modeling research are proceeded for connecting and interacting between virtual resources in cyberspace and physical devices/processes [7].

Current cyberspace identifiers are not able to present any physical location information. Although physical resources and cyber resources should be tightly coupled for advancing IoT applications, there is no human-friendly cyber-physical identification system.

There are several ways about discovering and connecting devices such as UPnP (Universal Plug and Play) [14] or DLNA (Digital Living Network Alliance). However, conventional methods use broadcasting beacon messages using specific network protocols for discovering devices/services, so it is hard to use as location identifier because protocols depend on their own specifications. Thus, they provide application based approaches, which actual users are not able to easily recognize.

3 Use Case and Motivating for Cyber-Physical ID/Location Mapping

As mentioned before, there is no method to acquire the representative service or information of a certain location in the IoT environment because mapping between physical address and cyber ID has been decoupled.

In this section, the usage and demand of Cyber-Physical ID/Location mapping (CPID) will be discussed based on two use cases of real service environment. Furthermore, the requirements of CPID are introduced.

3.1 Use Cases

3.1.1 Disaster network

The number of natural disaster is increasing, and its damage gets bigger and bigger, so the needs for standard of disaster network are dramatically increased. As response, ITU-T established the needs and requirements of disaster network in activities on "Focus Group on Disaster Relief Systems, Network Resilience and Recovery" [15]. In ITU-T Disaster-Relief (DR) requirement, it is said that a use of a wide variety of terminals and communication channels is a novel kind of method to early-disaster alert in disaster relief system. In previous disaster network, old-fashioned communication methods are used, such as radio, TV broadcast and wired telephone. However, in future disaster network, various kinds of devices would serve the disaster relief service, as wireless communication and digital signage technologies get evolved. On the network side, utilizing the NFV scheme, detouring network configurations can be much easily constructed. Moreover, these kinds of devices would provide area-specific or user-specific information to the user in disaster network [16].

When a disaster occurs, a wireless communication network might be congested by multiple coincident calls. In spite of this network status, an early alert message should be delivered to the users in the disaster area as soon as possible. Moreover, digital signage devices in bus station, railway station and retail outlet should be able to show proper messages for its area [17]. To guarantee these services trustworthy, it is required to be a public and easy-to-process cyber-physical ID mapping. A disaster network scenario is composed of following process - in the scenario, a fire in a public park and its remedy will be treated.

1) Observing location

A user in a public park sees a fire and reports it to the public park center or near fire station. The fire point gets recognized as a form of CPID based on representative point in the user's terminal and gets reported to the public park center. With spatial information in the report, the public park center is able to identify the fire location and its related resources, and make a proper decision to deal the disaster.

2) Broadcast information

The public park center decided to broadcast the fire information to users in the public park. The information of fire point is delivered as a form of CPID, so that it could be shown in proper form in user's mobile terminal and digital signage. The user is able to recognize the fire point by noticing the point that is included in the CPID form. This information could be applied to calculate refuge path or notify a safety zone.

3) Network disaster recovery

If there are any partial damage happens on the physical network around the public park, then re-configuration or detouring path can be straightforwardly build using NFV. By moving existing virtual appliance from the ruined place to other safe data center, associated application workload can be efficiently processed with minimum damage.

3.1.2 Smart city

A smart city environment has many scenarios for interpreting physical location to logical representation in IoT environment. In smart city or urban computing environment, a logical classification on the place gets complicated. Moreover, numerous heterogeneous services can be used in a certain area. To solve these issues, ITU-T discussed standardization issues on "Focus Group on Smart Sustainable Cites" [18]. In documents of smart sustainable cities overview and its standardization roadmap, components of services and ICT infrastructure are argued. Especially, in urban planning, intelligent building system and building information modelling, management of physical place and location with cyber system is essential. For this purpose, they argued that combination and harmony of related technologies are important.

For example, the location information scenario is one of representative scenarios for smart city environment. In previous, a visitor uses a map application or web search engine to acquire information of visiting location, or he/she can visit a web homepage of visiting location if it is served. However, the information of map application is dependent on its vendor, and it is not so detail for multi-story building or inside of building. Querying to web search engine is inconvenient solution to the user. If the visiting place provides some access points to get information, it would be better for the visitor. Still, it is incomplete solution since the access methods are dependent on its service vendor. Moreover, they just provide an isolated connection between physical location and cyber service. It is insufficient to apply to other mash-up services. A visitor scenario is composed of following process - in the scenario, a user visits a museum and uses a curator service by service discovering.

1) Observing location

A user visits a museum and recognizes his/her location using own mobile terminal. The mobile terminal acquires the location information from sensors or QR-code. The museum's location information get acquired in a form of CPID.

2) Service discovery and selection

The user searches services that are related to the museum by recognized locator. The location information in form of CPID is used as a parameter for searching. As results of service discovery, the user can find three services: information service of museum building, curator service for current display and information service for exhibition schedule. The user selects the curator service for current display.

When the user tries to search services related to current location, a searching method should not be depend on a certain vendor. It means that CPID needs to be public and platform-independent.

3.2 Requirements

From the use cases defined in Section 3.1, a list of requirements for CPID system has been identified. The detail and fined work of this list would be subject to future study. The requirements could be classified as three categories; presentation, sensibility and operability. Presentation requirements describe functional requirements to read, write and process CPID by computer or human. Sensibility requirements describe functional requirements for CPID to be easily sensible or readable by computer or human. Finally, operability requirements describe requirements of operations that ensure trustworthy services. The list of requirements is followed:

1) Presentation

- CPID needs to have a hierarchical structure. It makes easy to logically understand for human and provides advantages on sorting, searching and filtering process for computer.
- CPID presentation structure needs to be scalable and flexible, since logical classification on place could be modified by temporal, cultural or national reasons.
- CPID needs to be human-readable and have a clear unit of presentation.

2) Sensibility

- CPID needs to provide a trustworthy method to recognize user's location to provide area-specific or user-specific service.
- CPID system needs to provide a trustworthy method to recognize location of real estates or fixed devices like digital signage.

3) Operability

- CPID needs to provide users to make a data request or receive data via CPID, which is based on current nation's physical address system.
- CPID system needs to be trustfully operated by multi-stakeholders. For this purpose, CPID system needs to be defined as global standards.
- CPID information needs to be independent from application services.

- If new devices are connected to the network and get new cyberspace addresses, then CPID system needs to provide a method to acquire CPID for those cyberspace addresses. Furthermore, CPID system needs to provide a method of new devices discovery.
- The physical address needs to be trustworthy, which is acquired from CPID system.

From the list of requirements for CPID-based ID/location resolution, trustworthiness comes to the fore of important service component for public services and an ICT infrastructure. For example, service structure like Google's Physical Web [19] could be dependent to its object ID resolver vendor, so that it could not guarantee a public trust. For an ICT infrastructure as public goods or social overhead capital or network resources, it needs to be managed by governments or public organizations. Moreover, it needs to establish a global standards to make a trustworthy system since the ICT infrastructure could be inter-operated with multiple countries.

3.3 Trustworthy Cyber-Physical ID/Location Mapping System

Currently, object-oriented ID systems and cyber-physical ID/location mappings vary on nations, vendors and service domains. This situation would make it difficult to use location information, and make its domain be localized. Moreover, it would cause an additional cost for cross-domain conversion and a confidence problem among stakeholders in B2C/B2B transaction.

As shown in Figure 1, to provide trustworthy applications, the location information from physical things network and identifier from cyber

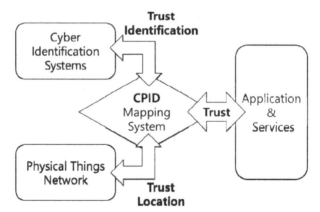

Figure 1   A concept of trustworthy cyber-physical ID/Location mapping system.

identification system should be trustworthy. A new cyber-physical ID/Location mapping system is needed to ensures trust location and trust identification, which includes location information explicitly for public goods in IoT environment, to make it easy to be recognized and used. Also, publicity and trustworthiness need to be ensured for this ID-based location mapping.

Structure that coupled with ID/location might have problems with massive routing or using in mobile environment [20]. However, a Cyber-Physical ID/Location (CPID) mapping system is needed for governance on public goods managed by location-based ID.

4 Standardization Activities and Trustworthy Cyber/Physical ID

In this section, we review current standardization activities about cyber ID and physical location.

4.1 Cyber Identification Standards

ITU-T Joint Coordination Activity for Identity Management (JCA-IdM) group has been established to standardize cyber identification system [21]. JCA-IdM also collaborates with various standardization groups such as ISO/IEC, ETSI, IETF, and so on.

The activities of JCA-IdM are based on ITU-T X.1252 document that provides a collection of terms and definitions used in identity management [22]. JCA-IdM has been discussed cyber identification systems, however, it is not considered that binding with location information or cyber-physical link.

The limitation of current cyber ID system is analyzed in ETSI GS INS 006 v.1.1.1 [23]. This document provides gap analysis for global discovery mechanism of identifier, provides and capabilities which is based on the assumption that the information required to provide a service is not available within a single service provider and must be dynamically discovered. Its main purpose is to investigate the current landscape on the IdM area and evaluate if there is a need for such a discovery mechanism, or whether this can be covered by existing solutions. In general, discovery of identity data across domains is realized with two different ways:

- Federated model: A service defined by a group of network entities which participate in a federation. Identity data are registered in the service and can be provided to all the participants of the group. The location of the discovery service and the protocol for exchanging messages is static and known to the participants of the group.

- User-centric model: By using an identifier of this format, a user directly points to a network point that holds identity information about itself. This location may hold information for only one profile of the user or for many profiles.

However, both approaches provide limited discovery of user's identity information. For the federated model, only the identity data, which exist within the federation of providers can be discovered. Information outside the federation cannot be discovered. For the user-centric model, the use of a specific predefined format instantly excludes the discovery of identity data from providers that are not familiar with it. Even though the adoption of a globally accepted identifier would solve major identity issues, which seems to be inapplicable.

4.2 Physical Location Standards

To utilize geo-location information in various area, it is needed that standards for aggregating, processing, and distributing geo-location data.

Open Geospatial Consortium (OGC) mainly develop and implement standards for geospatial content and services. OpenGSI Location Services (OpenLS) defines core services, their access and abstract data types which form together a framework for an open service platform, the so called GeoMobility server [24]. OpenLS provides four core services: Directory Service, Gateway Service, Geocoder Service, and Presentation Service. These services are based on GPS technology.

OGC has identified the need for standardized interfaces for sensors in the Web of Things (WoT). The Sensor Web interface for IoT SWG aims to develop such a standard based on existing WoT portals with consideration of the existing OGC Sensor Web Enablement (SWE) standards [25, 26]. The importance of location information and sensor observations to the IoT has been recognized.

OGC Sensor Web Enablement (SWE) has been established for building a unique and revolutionary framework of open standards for exploiting Web-connected sensors and sensor systems of all types: flood gauges, air pollution monitors, stress gauges on bridges, mobile heart monitors, Webcams, satellite-borne earth imaging devices and countless other sensors and sensor systems. SWE standards are the only ones that focus on the content of sensor information and on making the sensor observations useful to end user applications. SWE standards allow users to assessment the fitness for use of observations and to allow accurate processing on the sensed information to create derived information suitable to the user needs.

In much the same way that HTML and HTTP standards enabled the exchange of any type of information on the Web, the OGC SWE standards enable the discovery of sensors and corresponding observations, exchange, and processing of sensor observations, as well as the tasking of sensors and sensor systems.

Geographic information is gathered and used by OGC standards, however, geo-location information are not matched with cyber identifiers. Therefore, a new identification structure should be discussed for binding both cyber-physical ID/Location information.

5 Conclusion

As an age of the Internet of Things comes, cyberspace and physical world would be tightly and closely coupled. It makes that physical objects and locations are connected to cyberspace service, which usages are vary from a level of sensor network like smart home, smart factories to GPS-based navigation system, location-based services and disaster relief network using NFV. These kinds of connection start with mapping physical location and information to the cyberspace service.

However, trustworthiness on IoT and NFV services would be ensured when users or services could recognize and rely on services which are coupled with precise cyber-physical information. Therefore, new identification system is needed, which includes location information explicitly for public goods in IoT environment, to make it easy to be recognized and used. Also, publicity and trustworthiness need to be ensured for this ID-based location mapping.

Currently, related issues are under discussion in various standard organizations. To develop standards for trust ICT infrastructure, it is needed to develop global CPID system by discussing ID/Location issues in global cyber identification or geolocation standardization groups.

Acknowledgment

This work was supported by the ICT R&D program of MSIP/IITP. [R0190-15-2027, Development of TII (Trusted Information Infrastructure) S/W Framework for Realizing Trustworthy IoT Eco-system].

References

[1] Global Conference on CyberSpace 2013. (2013). *Seoul Framework for and Commitment to Open and Secure Cyberspace.* Available at: http://www.mofat.go.kr/english/visa/images/res/SeoulFramework.pdf

[2] World Summit on the Information Society. (2015). *WSIS Forum 2015: Outcome Document.* Available at: http://www.itu.int/net4/wsis/forum/20 15/Content/doc/outcomes/WSISForum2015_OutcomeDocument_Forum Track.pdf

[3] ITU-T Study Group 13. (2015). *Trust Correspondence Group.* Available at: http://www.itu.int/en/ITU-T/studygroups/2013-2016/13/Pages/corres pondence.aspx

[4] Zheng Y., Zhangc, P., and Vasilakosd, A. V. (2014). A survey on trust management for internet of things. *J. Netw. Comput. Appl.* 42, 120–134.

[5] Roussos, G. (2008). *Networked RFID: Systems, Software and Services.* London: Springer SMB.

[6] Roussos, G., and Chartier, P. (2011). Scalable ID/Locator Resolution for the IoT," in *Proceedings of the 2011 IEEE International Conferences on Internet of Things, and Cyber, Physical and Social Computing.*

[7] Lee, E. A. (2010). "CPS Foundation," in *Proceedings of the 47th Design Automation Conference*, Anaheim, CA, 737–742.

[8] ISO. (1999). *Codes for the representation of names of countries and their subdivisions*, ISO 3166.

[9] ITU-T. (2010). *The International Public Telecommunication Numbering Plan*, ITU-T E.164.

[10] RFC. (1998). *Internet Protocl Version 6 (IPv6) Specification*, IETF RFC 2460.

[11] Hui, J., and Culler, D. (2009). "6LoWPAN: incorporating IEEE 802.15.4 into the IP architecture," *Internet Protocol for Smart Objects (IPSO) Alliance White paper.* Available at: http://www.ipso-alliance.org/wp-content/media/6lowpan.pdf

[12] Raggett, D. (2015). *An Introduction to Web of Things Framework.* Available at: http://www.w3.org/2015/04/w3c-wot-framework-munich-2015.pdf

[13] RFC. (2002). *Report from the Joint W3C/IETF URI Planning Interest Group: Uniform Resource Identifiers (URIs), URLs, and Uniform Resource Names (URNs): Clarifications and Recommendations*, IETF RFC 3305.

[14] ISO. (2011). *Information Technology - UPnP Device Architecture*, ISO/IEC 29341.

[15] ITU-T FG-DR&NRR. (2014). *ITU-T Focus Group on Disaster Relief Systems, Network Resilience and Recovery.* Available at: http://www.itu.int/en/ITU-T/focusgroups/drnrr/Pages/default.aspx

[16] ITU-T FG-DR&NRR. (2014). *Requirements for Disaster Reflief System.* Available at: http://www.itu.int/en/ITU-T/focusgroups/drnrr/Documents/fg-drnrr-tech-rep-2014-5-DR_requirement.pdf

[17] ITU-T FG-DR&NRR. (2014), *Promising Technologies and Use Cases - Part IV and V.* Available at: http://www.itu.int/en/ITU-T/focusgroups/drnrr/Documents/fg-drnrr-tech-rep-2014-2-2-Framework-usecase-part-4-5.pdf

[18] ITU-T FG-SCC. (2015). *ITU-T Focus Group on Smart Sustainable Cities.* Available at: http://www.itu.int/en/ITU-T/focusgroups/ssc/Pages/defaul t.aspx

[19] Google. (2015). *Google Physical Web.* Available at: https://google.github.io/physical-web/

[20] RFC. (2007). *Report from the IAB Workshop on Routing and Addressing*, IETF RFC 4984.

[21] ITU-T JCA-IdM. (2015). *ITU-T Joint Coordination Activity for Identity Management.* Available: http://www.itu.int/en/ITU-T/jca/idm/Pages/def ault.aspx

[22] ITU-T. (2010). *Baseline Identity Management Terms and Definitions*, ITU-T X.1252.

[23] GS INS. (2011). *Identity and Access Management for Networks and Services; Study to Identify the Need for a Global, Distributed Discovery Mechanism*, ETSI GS INS 006.

[24] OGC. (2008). *OpenGIS Location Service (OpenLS) Implementation Specification: Core Services*, OGC 07-074.

[25] OGC. (2011). *Sensor Web Enablement.* Available at: http://www.opengeo spatial.org/domain/swe

[26] OGC. (2012). *SensorThings Standards Working Group.* Available at: http://www.opengeospatial.org/projects/groups/sweiotswg

Biographies

H. Oh received his B.S. and M.S. degree from Korea Advanced Institute of Science and Technology (KAIST) in 2012 and 2014, respectively. Currently, he is a Ph.D. student in KAIST. His current research interests include energy efficiency in heterogeneous access networks, Internet of things, Web of objects, and user interface/user experience on Web.

S. Ahn is a Ph.D candidate in electrical engineering (Korea Advanced Institute of Science and Technology, Daejeon, Republic of Korea). He received his Bachelor's degree in computer science from KAIST. He received his Master of engineering degree in electrical engineering from KAIST in 2010. His research interests include web engineering, insulated networking mobile cloud computing and services in the Internet of Things.

J. Yang (S'05) received M.S. in computer science from InJe University in 2005 and HERIT Inc. in 2008 and currently he is Ph.D. candidate student in Korea Advanced Institute of Science and Technology (KAIST). His main research interests include next generation network, multimedia streaming issues, and IoT.

J. K. Choi (M'88-SM'00) received the B.Sc. (Eng.) from Seoul National University in electronics engineering, Seoul, Korea in 1982, and M.Sc (Eng.) and Ph.D degree in 1985 and 1988, respectively, in electronics engineering from Korea Advanced Institute of Science and Technology (KAIST). From June 1986 until December 1997, he was with the Electronics and Telecommunication Research Institute (ETRI). In January 1998, he joined the Information and Communications University (ICU), Dae jeon, Korea as Professor. At year 2009, he moves to Korea Advanced Institute of Science and Technology (KAIST) as Professor. He is a Senior Member of IEEE, the executive member of The Institute of Electronics Engineers of Korea (IEEK), Editor Board of Member of Korea Information Processing Society (KIPS), Life member of Korea Institute of Communication Science (KICS).

A Study on the Network Resource Openness with Software Networking toward the Development of the Web Technology

Jinhong Yang[1], Hyojin Park[2] and Jun Kyun Choi[2]

[1]*Information and Communications Department, Korea Advanced Institute of Science and Technology, 291, Daehak-ro, Yuseong-gu, Daejeon 34141*
[2]*School of Electrical Engineering, KAIST, Republic of Korea*
E-Mail: {jinhong.yang; gaiaphj}@kaist.ac.kr; jkchoi@ee.kaist.ac.kr

Received 22 January 2016; Accepted 2 March 2016;
Published 15 March 2016

Abstract

The continuous development of the network and the growth of the Internet and the Web are festinating. The improvement in network softwarization technologies represented by the SDN and NFV made the dedicated legacy network resources into more elastic to be utilized. These progresses realize the accessibility on the customized network resource utilization for web services, overcoming traditional best effort network environment. HTTP based application services are extended from multimedia to IoT with fast evolution of Web technology. In this study, network resource open trends are explored in softwarization networking technology perspective, and describe what will be the future of network environment, step-by-step. Especially, the process of change is focused from previous individual communication service environment such as voice, data to unified communication environment.

This work was supported by Institute for Information & communications Technology Promotion (IITP) grant funded by the Korea government (MSIP) (No. B0101-15-0331, Development of Multi-Screen Adaptive Autonomous Smart Media Technology)

Journal of Software Networking, 19–36.
doi: 10.13052/jsn2445-9739.20161.002

Keywords: Network resource, Software Defined Network, Service Openness, Web Technology, WebRTC.

1 Introduction

Evolution of the telecommunication network's architecture has made their network resources to become more available and utilizable for various network stakeholders. As the theme of the network, Internet, and Web evolves to objectify their resources and operate in software-based frame to have openness so that can have more opportunity to be used in the new environments, the network architecture and solutions have also been developed to support such changes [1, 2]. The progress of such evolution can be described in resource accessibility-based network architectural point of view by their composition of the compounding components. In early 1990s and later, the network components were tightly coupled and connected within the network architecture with monolithic Pre-Software Oriented Architecture (SOA) like composition. Later, in 2000s, the concept of SOA was widely adapted in IT industry and the rapid expansion of Internet has brought the discussions on the reusability and openness of the network resources. Accordingly, the traditional SOA technology has been adapted to the network architecture and the compounding network components became much loosely coupled [3]. After that, with the emergence of the All IP concept, the network has evolved into the Next Generation Network (NGN) architecture and the network components are forming more decoupled compositions. Recently, such trend is more specified under the name of the Software Defined Network (SDN) and Network Function Virtualization (NFV). As the more techniques and schemes are developed for the network components to be virtualized and build with software-based composition, even the static provisioning based legacy network resources became dynamically available for services in just in time manner [4].

To this change, the rapidly enhancing improvement of the Internet and the Web has hugely affected. The changes from circuit based data transmission to the packet-based data transmission brought by the global adoption of the IP based internet protocol diminished the burden of linkage among network resources. The overriding penetration of the Web made the HyperText Transfer Protocol (HTTP) and the client-server based service architecture enough for almost every applications and services over browsers. Such environmental domination of prominent technology has simplified the network transport layer and adopting the concept of Web 2.0 based service architecture enabled the network components become objectified.

Nowadays, Web is not just a useful function over the telecommunication network, but an essential tool for our daily lives. As a point of contact to the cyber and online services, Web supports data portal, information search, social media service, rich multimedia content display, and so on. Recently, by utilizing the Web architectures, Web service, and Web of Device technologies, Web expands its service territory to even 'things' [5]. By doing so, Web performs the role of global data platform. The role as a global data platform is not wholly come from the development of the Web technology itself or the Web based service technology. Now, the network operators have clear role in the Information and Communications Technology (ICT) industry; as a physical and logical data infrastructure provider and manager, conducts OAM (operations, administration and maintenance) like the electric power corporations or highway corporations. It will keep the stable revenue out of doing what they have been doing. However, to get the new cash cow, it is important to looking after the new business opportunity as the Web grows in its volume and evolves in its capability while generating new service trends and requirements for existing infrastructures.

This papers studies on how to adopt the evolving trend of the network resource and service openness led by the development of the Web technology in network operator's point of view. Because the Web is the subject who requires to change the object of the resource openness from the network providers to the internet service providers. For that, this paper is composed as follows: In Chapter 2, the development of network resource open service interfaces and the state-of-the-art Web technologies are introduced as background technology. Then, the changes in the telecommunication service environment surrounding network are discussed in Chapter 3. After that, in the Chapter 4, as an use case, the deployment strategies for voice telephony over Web is described in three stages and concludes in Chapter 5.

2 Development of Network Open Service Interfaces and Web Technology

2.1 History of Network Open Service Interfaces and APIs

The network operators have expected to utilize their network resources in various ways so that can secure multiple sources of revenue. In that manner, the IT industry speedily upgraded their environmental conditions from the

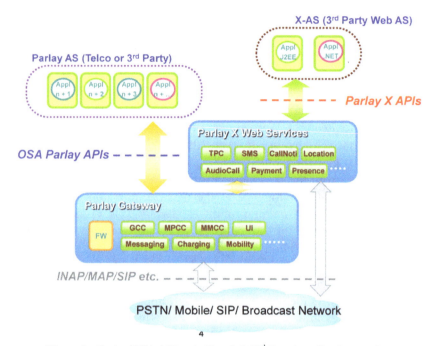

Figure 1   Parlay/OSA API and eXtended (3rd Party) application service.

beginning of the 21st century. They tried to provide the network based services to have the form of easily created and deployed, executed and managed features in accordance with the network service user and provider's requirements. That intention is revealed from Parlay/OSA on the intelligence network environment to the Web services over the Web 2.0 environment and the Service Delivery Platform with the Service Oriented Architecture concept [6–8]. Figure 1 shows the concept of the Parlay/OSA based network resource utilization and the Parlay X based connectivity with 3rd parties. Figure 2 demonstrates the reference architecture of the network service integration and delivery environment, which has developed in the ITU-T's standard recommendation Y.2240, "Requirements and capabilities for NGN Service Integration and Delivery Environment (NGN-SIDE)". The standard supports to open the following network resources and functions over the Next Generation Network (NGN) environment [9].

- Integration of resources from different domains over NGN (e.g. telecom domain (fixed and mobile networks), broadcast domain, internet domain, content provider domain etc.)

- Adaptation, including abstraction and virtualization, of resources from different domains
- Resource brokering for mediation among applications and resources
- Application development environment for application developers
- Different service interfaces across ANI, UNI, SNI and NNI for exposure of NGN-SIDE capabilities and access to resources in different domains
- Mechanisms for support of diverse applications, including cloud, machine to machine, and ubiquitous sensor network applications
- Mechanisms for support of applications making usage of context based information
- Mechanisms for content management

Through the Service Delivery Platform (SDP), by offering each network resources in the form of service enablers, the network service operators have

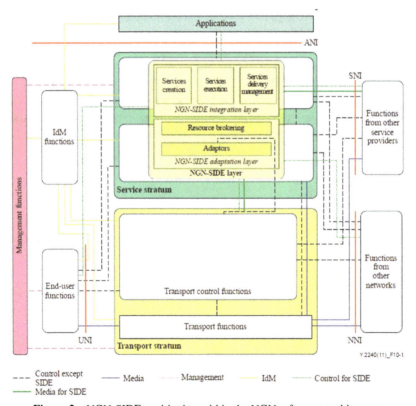

Figure 2 NGN-SIDE positioning within the NGN reference architecture.

provided the network resource users to be able to utilize the network resources via Web service or Application Programming Interfaces (APIs).

In addition, recently, as the software-based network resource management and operation technologies, called SDN and NFV are introduced, such evolution toward network resource and service openness is even accelerating. Especially, the major features of the SDN like Programmability, Granularity, and Elasticity turned the application-awareness based service available, and embraced the opportunity of providing the capabilities on the network transport stratum, which were both not possible to be supported in the legacy network resource opening technologies [10].

2.2 Web Technology

As the Internet based services uses the HTTP for their services over the Web, Web is expanding its core technology from a markup-based content rendering into devices, networks, and applied services. As described below, Web has kept extending to a large convergence of software.

When it comes to the latest Web technology, W3C explains the openness of the Web as below [11]: W3C standards define an Open Web Platform for application development that has the unprecedented potential to enable developers to build rich interactive experiences, powered by vast data stores, which are available on any device. Among the several Web technologies, the HTTP 2.0 and the WebRTC plays to major role to contribute for the network resource utilization.

2.2.1 HTTP 2.0

Version 0.9 of HTTP was announced at 1991, HTTP/1.0 at 1996, and 1.1 at 1999. After HTTP/1.1, there were no changes on the HTTP over 10 years. However, today's web pages are about 20 times larger than 1990's in the aspects of web pages' size and number of requests and raised few issues [12].

First of all, HTTP/1.0 allowed only one request at a time on a given TCP connection. HTTP/1.1 added request pipelining, but this only addressed partially of the request concurrency problem and still has an HOL (head-of-line) blocking problem. Therefore, to achieve concurrency while reducing latency, HTTP/1.0 and HTTP/1.1 clients who desirably want to make many requests need to be able to use multiple connections to a server. Furthermore, HTTP header fields are often repetitive and lengthy. As everyone can forecast, it caused unnecessary network overhead seriously.

The features of HTTP/2.0 to handle these issues mentioned before and to achieve better performance with low latency are as below.

- HTTP header compression
- Binary protocol
- Multiplexing
- Full-duplex interleaving and Stream priority
- Server push

2.2.2 WebRTC

According to the introduction of the 'webrtc.org', WebRTC is a free, open project that provides browsers and mobile applications with Real-Time Communications (RTC) capabilities via simple APIs.

WebRTC offers web application developers the ability to write rich, real-time multimedia applications (think video chat) on the web, without requiring plugins, downloads or installs. Its purpose is to help build a strong RTC platform that works across multiple web browsers, across multiple platforms.

Figure 3 is the overall architecture of WebRTC [13].

WebRTC has three APIs. The first one is MediaStream API. It represents synchronized streams of media. For example, a stream taken from camera and microphone input has synchronized video and audio tracks. Second is

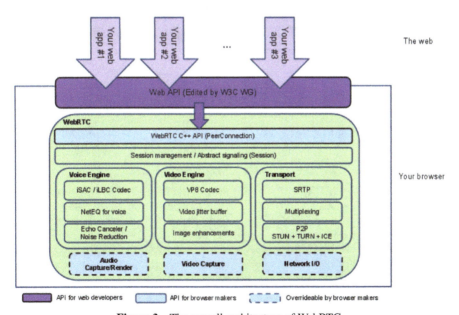

Figure 3   The overall architecture of WebRTC.

RTCPeerConnection. It is the WebRTC component that handles stable and efficient communication of streaming data between peers. RTCPeerConnection offers easier way to access to core function without numerous complexities from underlying system. The last one is TCDataChannel. As well as audio and video, WebRTC supports real-time communication for other types of data. The RTCDataChannel API enables peer-to-peer exchange of arbitrary data, with low latency and high throughput. It can be used for gaming, remote desktop applications, real-time text chat, file transfer, and decentralized networks.

3 Changes on Service Environment Surrounding Network

Telco first started wired voice service on PSTN and have gradually expanded their business area to wired Internet service and mobile-based voice and data services. With the beginning of the new millennium, telco's network environment was transformed to All IP network and wired broadband. Telco then have broadened its service upon this technology advancement: from the conventional telecommunication infrastructure service to contents and media, and infra-optimized utility services. Telco got off from the simple business model and now seek to widen its value chain by all-out mobilization of their capability.

In the Telco X.0 environment, Telcos' main services are as follows: wire and wireless convergence and broadband services based on All IP Network, smart services based on contents and media, and utility services backed up by Telco infrastructure as depicted in Figure 4.

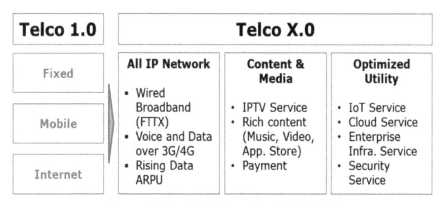

Figure 4   Value Chain Expansion via Telco X.0.

3.1 Optimized Utility

As the service environment has been evolved from H/W focused from S/W oriented, different services utilizing the resources of Telco are introduced in the market. Cloud service and enterprise infrastructure are one type of those services, and it focuses more on providing computing resources over the network infrastructure, rather than on supplying the network itself. The end-to-end quality of network, however, needs to be assured to make this service viable. As software defined network infrastructure became widespread recently, networks are virtualized through SDN, and these resources shared and provided in the form of IaaS (Infrastructure As a Service) or PaaS (Platform As a Service).

Moreover, recent-soaring IoT (Internet of Things) services are projected to broaden the service range by connecting devices in a more accessible and universal way, which was limited in previous M2M (machine to machine) utilization.

The main reason that the convergence is called for between the Web and Telco is that the conventional models of both are in a counter position; they accordingly come to be located in a complementary relationship when it comes to provide new business zone to telecommunication service providers in Figure 5.

For Telco, abundant Internet and Web technology is required to support a high-level of Internet application for Cloud computing, future Web, and Cloud service. This Web-Telco convergence will enable Telco to take the role of major supplier of global services for both enterprises and personal users as depicted in Figure 6.

Strategic partnership between the Web and Telecommunication technology is needed, for both technologies share significant similarities and has a complementary dynamics at the same time. What Telco X.0 technology should

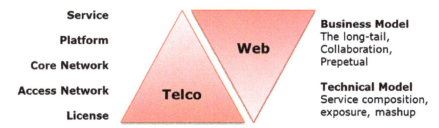

Figure 5 The opposite relationship between the conventional Telco model and the Web model.

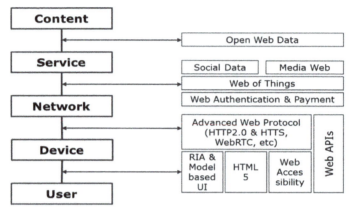

Figure 6   Overview of Web technology stack on.

pursue are, especially in the perspective of the evolution from the traditional Telco technology, two points: (1) the collaboration of content provider, who is the key player of Web 2.0, and network/communication technology provider and (2) the discovery of new business model.

Considering the major changes Telco X.0 would go through, Web technology is a must for conventional network providers, since the convergence between Web and Telecommunication tech is expected to speed up upon the

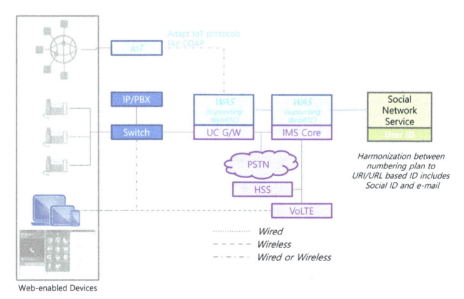

Figure 7   UCWeb evolution: stage 1 for Web-based user interfaces and unified gateway.

complementary relationship of them. The matter is, however, who pays and who earns the money, which is difficult to be realized without the connectivity between Telco and the Web.

3.2 Everything as a Service

Simpler and more cost-efficient methods for doing Internet business have shown a rapid growth for the last few years. The market is moving toward an As-a-Service delivery model that provides plug-in, scalable, consumption-based business services that deliver the business outcomes that every organization demands – increased revenue or decreased costs.

Represented as "X-as-a-Service", this model provides technical methods to utilize all the resources on-demand, real-time, and usage-based; previously, service providers had to build and manage whatever needed to provide a certain internet service.

For example, a start-up company that wants to try out a new internet business in the market formerly would have to build its own server and network, appoint corresponding managers for each task, and develop own software for its own. Nowadays, in contrast, it can use whatever resources it needs and pay the only amount accordingly, regarding the whole on-demand utility of server and network, the corresponding management, and the software application working on those infrastructures.

4 Using Network Resources for Unified Communications over Web

With viewpoint of telecommunication industry, the unified communication service over web (UCWeb) is the wider extension from voice over IP (VoIP) to adapt web environments. The VoIP is mainly focusing on communication protocol to deliver voice telephony service. But, the UCWeb is utilizing web technologies to deliver voice/video, IoT and many other Web-based services.

The clear feature of UCWeb is that the end users keep the web environment such as HTML and HTTP based on URL/URI. It assumes that the user device uses web-enabled screen and all the end user services are applicable through the web. The communication protocol is based on HTTP or HTTPS protocol. The display screen of end user is running on HTML.

With graceful evolution toward future web convergence, the UCWeb has to maximize the interoperability with the existing VoIP technologies. The UCWeb may use the existing VoIP protocols such as Session Initiation

Protocol (SIP), Session Description Protocol (SDP), Real-time Transport Protocol (RTP), and Real-time Transport Control Protocol (RTCP), etc. In this case, The UCWeb can use the existing numbering scheme running on VoIP and 3G/LTE network.

According to the UCWeb evolution, the following stage from the existing wireline VoIP network may be considered.

4.1 Using Unified Communication Gateway

First stage is UC Gateway connecting to Web Server in Figure 8.

- The end user has two options to make voice call: one is via VoIP gateway and the others is directly connected to UCWeb server by using WebRTC. The end user can directly access the web services through WAS (web application server). The WAS offers various types of web applications by combining location and presence information. Special messaging service may be added.
- The IMS core may be included at the web server. The VoIP user may call to the customer of social network service and instance messaging services with help of WAS. Vice versa, the social customers may call the VoIP users using Social ID.
- The Internet-of-Thing (IoT) devices may be linked to web server. When the human may need to control and manage IoT devices via signalling,

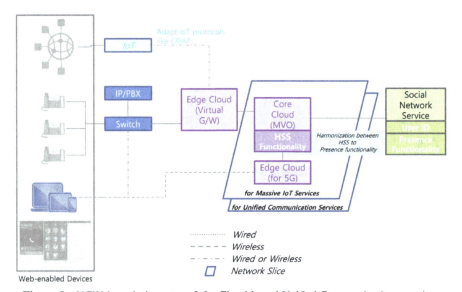

Figure 8   UCWeb evolution: stage 2 for Cloud-based Unified Communications services.

the WebRTC protocol may be used, which is especially for IoT devices equipped with camera and speaker.

- With the extension of the existing numbering plan, the web-accessible user identification such as web ID (URL/URI), email address, and social network service SNS identifications can be used. The user ID of social network service has to be recognized by VoIP customers. The SNS users may search VoIP customers to make a call.

- The direct connection from end user to web server is relying on web standards like HTTP or Constrained Application Protocol (CoAP) etc. The WebRTC can be used to connect the web server. The WebRTC may utilize VoIP protocol to make a call.

- The emergency call or group call may also support with two options described above; via VoIP gateway or directly connect to web server. The group community call of social network services is very useful for simultaneously notification or announcement of emergency to all the members.

- The charging and billing server may connect to the web server, which may replace HSS at VoIP and 3G/LTE network in future. By combining with other web applications, various charging and billing options can be considered.

- The call-based intelligence services (e.g., call center services) can be directly connected to web server by using web APIs or the cloud concept of XaaS (anything as a service). It provides the add-on features of web services and web applications.

4.2 Cloud-based Unified Communication Gateway

The next stage is the Cloud-based Unified Communication Gateway replacing UC Gateway as shown in Figure 8. The roles and key features of each component in the figure are as followings:

- The core cloud (e.g., web server with presence) is a center for all the services and applications provided by Web technology. The signals in between users and devices are transmitted by HTTP 2.0 and the media information are exchanged by WebRTC.

- The end users may connect to the regional web presence server with/without the intermediate gateway and proxy system. To provide the global scalability of web presence server, the cloud-based distributed configurations of web servers are necessary to reach the billions of end users.

- The cloud platform reaches to end users with wireline switch and/or wireless access gateway including RNC and BSC. The concept of network functional virtualization (NFV) may be used to maximize channel efficiency and avoid congestion at access network area.
- According to the migration stage, the access cloud platform may include the unified communications gateway. But, alternative signalling solutions may be chosen depending on applications.
- The mobility support is relying on the available wireline and wireless technologies. But the current 3G/LTE tunnelling and traditional VoIP protocol will be eventually upgraded. The end user device like smartphone may have some intelligence on mobility by using buffer and intelligent movement detection. Without complicate handover technique, smartphone can help the seamless connectivity by allowing some duplication or redundancy of packet streams while user moves.
- For charging and billing options, a large variety of business opportunities including sharing economy may be coming at near future. The traditional home subscriber system (HSS) may be upgraded in order to offer plentiful business solutions.

5 Conclusion

The network resource and service openness and softwarization have been progressively developed in both technology and architectural manner. Recently, it is showing even widened coverage which is from the network application level (L7) to the data link level (L2) via the state-of-the-are researches on SDN and NFV. Along with that, the Web technology has also been steadily developed to utilize the new resources. Evolved from the early form of the link among web pages, current Web technology is applied in a wide range of services: sensor/device, service/application, and cloud. As with the giant leap of Web-based media (social media, video portal, etc.), the Web is now essential for people's daily life.

This paper looks out the evolution of the two technologies and provides a blueprint on the enhanced communication service architecture using WebRTC over the SDN and NFV based upcoming network environment. The network, the core asset of Telco which gives the grounds for its business, will evolve to be functionalized via SDN/NFV and if so, software will be at the center of network control. Network provider may amplify the Web leverage by utilizing resources in a lighter and smaller unit, which is achievable with the adoption of microservice-based software architecture and DevOps. Therefore, further study on this area is required to enhance its cost-effective applicability in current network.

References

[1] Chung, J. Y., Jeong, T., Yoo, J.-H., Hong, J. W.-K. (2015). "Define interfaces for network softwarization," in Proceedings of KICS, 870–871.

[2] Shin, M.-K., Nam, K.-H., Kim, H.-J. (2012). "Software-defined networking (SDN): a reference architecture and open APIs," in 2012 International Conference on ICT Convergence (ICTC), 360–361.

[3] Magedanz, T., Blum, N., and Dutkowski, S. (2007). Evolution of SOA concepts in telecommunications. *Computer* 40, 46–50.

[4] Aslan, M. and Matrawy, A. (2016). On the impact of network state collection on the performance of SDN applications. Commun. Lett. IEEE 20, 5–8.

[5] W3C *Web of Device*. Available at http://www.w3.org/standards/webofdevices/

[6] W3C *Web Service Architecture*. Available at https://www.w3.org/TR/ws-arch/

[7] Bennett, A. J., Grech, M. L. F., Unmehopa, M. R., and Vemuri, K. V. (2003). Service mediation standards. Bell Labs Tech. J. 7, 77–90.

[8] Qiao, X., Chen, J., Gu, P., and Li, Y. (2014). Opening up telecom networks with a lightweight web element service cloud for ordinary users in the web 2.0 era. IEEE Commun. Mag. 52, 127–133.

[9] ITU. (2011). *Requirements and Capabilities for NGN Service Integration And Delivery Environment (NGN-SIDE)*, ITU-T Recommendation Y.2240.

[10] Jarschel, M., Zinner, T., Hossfeld, T., Tran-Gia, P., and Kellerer, W. (2014). Interfaces, attributes, and use cases: A compass for SDN. IEEE Commun. Mag. 52, 210–217.

[11] W3C *Standards*. Available at http://www.w3.org/standards/

[12] Average Web Page Breaks 1600K. Available at http://www.websiteoptimization.com/speed/tweak/average-web-page/

[13] WebRTC. Available at http://webrtc.org

Biographies

J. Yang (S'05) received M.S. in computer science from InJe University in 2005 and HERIT Inc. in 2008 and currently he is Ph.D. candidate student in Korea Advanced Institute of Science and Technology (KAIST). His main research interests include next generation network, multimedia streaming issues, and future media.

H. Park (S'07) received M.S. in communications engineering from Information and Communications University in 2007 and currently, she is Ph.D. candidate student in Korea Advanced Institute of Science and Technology (KAIST). Her main research interests include IPTV, broadcast networks, multimedia streaming issues, and next generation network.

J. K. Choi (M'88-SM'00) received the B.Sc. (Eng.) from Seoul National University in electronics engineering, Seoul, Korea in 1982, and M.Sc. (Eng.) and Ph.D. degree in 1985 and 1988, respectively, in electronics engineering from Korea Advanced Institute of Science and Technology (KAIST). From June 1986 until December 1997, he was with the Electronics and Telecommunication Research Institute (ETRI). In January 1998, he joined the Information and Communications University (ICU), Dae jeon, Korea as Professor. At year 2009, he moves to Korea Advanced Institute of Science and Technology (KAIST) as Professor. He is a Senior Member of IEEE, the executive member of The Institute of Electronics Engineers of Korea (IEEK), Editor Board of Member of Korea Information Processing Society (KIPS), Life member of Korea Institute of Communication Science (KICS).

A Study on the IoT Instance Hosting in Edge Cloud and NFV

Jinhong Yang[1], Hyojin Park[1], SungKwan Jung[1]
and Hyuncheol Kim[2]

[1]*Dept. of Information and Communications,*
Korea Advanced Institute of Science and Technology, South Korea
[2]*Dept. of Computer Science, Namseoul University, South Korea*
E-mail: {jinhong.yang; gaiaphj,skjung}@kaist.ac.kr; hckim@nsu.ac.kr

Received 2 May 2016; Accepted 13 June 2016;
Published 8 July 2016

Abstract

Recently, as the number of IoT devices generate tremendous volume and variability of data. By the time data makes its way to the cloud for processing the chance to act on it might be disappeared. This paper presents a novel idea, proposes methods for virtualizing Internet of Things devices and hosting them on the home access point instead of relying on a cloud. For this, we design profile of the IoT gadgets and virtualize into JavaScript-based objects. Then, to execute and control the instances of the virtualized IoT objects on the wireless AP, a novel instance management method and their interfaces are designed. The implementation and performance test demonstrates the proposed system's stability and operability in the real wireless AP running environment. Adopting this platform feature over SDN/NFV infrastructure might be beneficial in to the IoT mashup service quality and its network efficiency.

Keywords: Internet of Things, Virtual Object, Instance Hosting, Edge Cloud.

1 Introduction

Recently, smart gadgets, sensors and appliances for personal healthcare and home automation are rapidly being deployed, which builds up an IoT (Internet

Journal of Software Networking, 37–52.
doi: 10.13052/jsn2445-9739.20161.003

of Things) environment. On the contrary to the conventional sensor market which is target to the industry or service providers, this emerging sensory device market is target to the individual consumers in personal and home environment. This implies that the number and the kind of sensory devices belong to an individual person are also being increasing.

In these days, there are several kinds of App-Enabled accessories [1] or smart-home gadgets and sensors which are the consumer IoT devices that can be bought and configured by end users themselves. To control and access these gadgets requires their own service applications in foams of mobile applications, web services or native embedded system services.

This dedicated service application based access and control approach requires the end consumer to be burdened down with searching a dedicate service application for each IoT gadget and configuring it properly. And this approach also makes it more difficult the service provider to provide an integrated control and service mashup among several IoT gadgets. Thus, providing a vertical service solution for individual IoT gadget has hindered the development of IoT eco-system [2, 3].

To deal with this problem, there are some approaches, such as a service-based interworking approach and a cloud-based service approach. In a service-based interworking approach, the individual service (mobile application or web service) for a specific IoT device supports the registration of other IoT devices and provides an interworking function among IoT devices within its service. And in a cloud-based service approach, each IoT device connects to the cloud platform, and a service on the cloud platform provide all the necessary IoT service and interworking functions [4]. Using these approaches enables some more enhanced mashup services in user environment, but has a limit that the pre-built IoT service channels can interwork with each other [5]. Thus, users still suffer the limitation that any combination of IoT gadgets cannot be mashed up as his/her needs.

In this paper, a method that enables interworking and collectively controlling various IoT gadgets on a Wi-Fi AP (Access Point) system is suggested to solve these problems. Using a Wi-Fi AP system as an integrated IoT gadget controller has several benefits. First, it doesn't require any additional device but only the existing Wi-Fi AP system. Second, the personal/private data that is created from IoT gadgets can be stored and managed within his home safely. Third, more instantaneous service response is possible on the contrary to the cloud platform based service which might suffer from the service latency or drop due to network conditions. And last, it might support legacy (non-IoT) devices due to the possibility of supporting various physical device interfaces. To achieve these benefits, we propose an optimized IoT gadget integration system and management scheme for its IoT service interworking feature with regarding the limited resources of Wi-Fi AP's and their embedded programming environment.

For this, a virtualization method to make a real world IoT device as a virtual object (VO) to be used in mashup environment is described, and the interface design and service management system architecture to create, execute and manage the created VOs on the in-Home Wi-Fi AP system is suggested. Also, the performance evaluation of the suggested platform is described to verify its intended operational behavior and its effect on the packet routing performance of Wi-Fi AP system by implementing the suggested platform on the real Wi-Fi AP. At last, the advantages and consideration points of making an IoT edge cloud over SDN/NFV infrastructure by extending the suggested IoT integrated management platform feature are described.

2 Related Works

The early stage IoT researches were mainly focused on how to connect physical things and devices to Internet. After that, the main research trend is how to utilize various connected IoT devices effectively. Thus, these researches are focus on the interworking various gadgets and the integration and analysis of their data. But, they are mainly beneficial to the large scale service providers, not to the end users who want mashup their own IoT gadgets as their needs. For example, Pachub (now, Xively) service only provides a simple triggering feature from IoT device data based on user-specified regular expression and doesn't provide any inter-device message exchange nor any additional service logic to make a mashup service.

The research on the virtualization of physical things could be classified the research on the device profile and interface and the research on the programming model for virtual object handling. First, for the description of device information, most of the researches adopt the standard model such as DPWS (Devices Profile for Web Services), and others develop its own profile model suit for their service model and environments. For the interface model, most of researches on Web of Things (WoT) utilize the SOAP (Simple Object Access Protocol) or HTTP-based REST (Representational State Transfer) structure to achieve message interoperability [6–8]. The early researches on the programming model were mainly conducted on the limited embedded system, such as TinoOS or Contiki. Thus, these models only supported the very limited programming model which required the device logic to be pre-built with the C programming language and to be loaded in the embedded running environments. Meanwhile, most of now-a-days researches adopt the

VM (Virtual Machine) based programming models [9, 10]. This VM based model makes the service providing structure to be more general one, but requires some more high performance device platforms.

The researches on the execution of virtualized objects have followed the similar trend with the development of the programming models: from the C-based object code native execution to VM based indirect execution. In addition, as the service expandability via interworking with existing web services gets main focus, there are several researches on the web based execution model along with the external web services [11, 12]. In this web based execution model, IoT gadgets expose their controls and information in RESTful foams via gateway devices with web servers, and the external web services can receive IoT gadget messages and execute their own service logic under own web environments [13, 14]. This execution model has the advantage that makes the development of virtualized object functionalities more easy due to the supports of several script based languages for web services, such as HTML, JavaScript, PHP, Python and etc. In addition to that, the web based execution model provides the user friend running and visualizing environments via web browser interfaces. But, most of the researches on the web based execution model utilized the gateway device only for the interfacing media and did not utilize it for service logic executions.

Though recent research approaches can provide several advantages, they also have some problems;

Under the script-based VO execution environments, there might be too frequent context switches for loading several instances, which requires somewhat more computing resource.

But, there might be a lack of all-time available computing resource for handling several VOs on network edge realm, especially in the home networks,

Thus, we suggest a novel system model that can describe the IoT gadgets as VOs, load and execute them with a capability of flexible IoT service mashup over the very limited computing resources.

3 IoT Instance Hosting Platform

First of all, these are important to make an IoT device in real world as a virtual object to be used in cyber world, and to equip an operating environment for executing the device in order to implementing and controlling IoT device at wireless internet router. In this Chapter, a methodology of device virtualization, connecting method between objects through an interface definition and structure design of instance manager for operating them are described.

3.1 Schema for Defining VOs

Virtual object is a thing which is defined to an object describing IoT device and functions provided from the device. It contains basic profile information of a physical device, technologies of network interface and operation logic of object for executing in virtual environment. In this paper, objects are defined separately according to a usage of virtual object; thing associated virtual object and mashup service virtual object. Thing associated virtual object is an object made for providing same functions in remote environment based on the message from IoT device. Therefore, basic information about a physical device (name, image information, explanation, etc.) and information of network interface for connecting (interface type, network address, port number, etc.) are described in XML. A logic of the physical device is also described in JavaScript. Table 1 shows a summary of main elements in thing associated virtual object.

Table 1   Example schema for thing associated virtual object

```
<!- - Omit the definition of Schema - ->
<!- -"Profile" element is used as a top element - ->
<xs:element name="profile">
 <!- - Virtual Object unique id - ->
 <xs:element name="profileId">
 <!- - Virtual Object version - ->
 <xs:element name="profileVersion">
 <!- - Basic information of virtual object - ->
<xs:element name="instanceDetailURL">
 <xs:element name="instanceDescription">
 <xs:element name="instanceIconURL">
 <xs:element name="instanceRunningPeriod">
 <!- - Connecting information of thing associated virtual object - ->
 <xs:element name="associatedThingInformation">
 <xs:element name="configurableValue">
 <!- - Information related to interface between virtual objects - ->
 <xs:element name="interfaceInformation">
  <xs:element name="oAuth">
  <xs:element name="feeder">
  <xs:element name="subscriber">
  <xs:element name="receiver">
  <xs:element name="controller">
 </xs:element>
 <!- - Describe the executing logic of virtual object in JavaScript - ->
 <xs:element name="instanceScript">
 </xs:element>
```

Table 2   Example schema on filter information for mashup service virtual object

```
<xs:element name="subscriber">
<xs:element name="filter">
<xs:attribute name="profileId"/>
 <xs:attribute name="feederId"/>
 <xs:attribute name="minVersion"/>
 <xs:attribute name="maxVersion"/>
</xs:element>
</xs:element>
```

Mashup service virtual object is an object which has an independent service logic regardless of direct connection to a device. This object can provide a mashup service combined thing associated virtual object and service logic. Mashup service virtual object contains an extra filter information for searching connectable virtual object. Table 2 shows that filter information has a unique ID of connectable object for searching other virtual object, 'profileID', 'feederID' and version information.

3.2 Interface for Message Exchange Between VOs

Virtual object works by 'instance' which is the type of executable program of defined object when it is operated in the environment. Thus, thing associated virtual object is executed by 'thing associated instance', mashup service virtual object is executed by 'mashup instance'. In this paper, five interfaces are defined for connecting instances easily; 'Feeder', 'Subscriber', 'Controller', 'Receiver', 'OAuth'. Figure 1 shows interface connecting structure between instances graphically on an instance manager designed for controlling and executing virtual objects at wireless internet router. It also describes a relationship that mashup instance gets a message from the thing associated instance which received data from each device and provides mashup service connecting through OAuth and outside web service.

At first, 'Feeder' interface is an interface of exposing information for providing data and message generated from IoT device to instance or outside service. Feeder interface works the form to store information for feeding to Message Queue of instance, not to deliver message directly.

Secondly, 'Subscriber' interface provides a function that instance receives information from Feeder interface of other instance or outside thing. Mashup service object can be subscribed about interface information matched to the condition of 'filter' element.

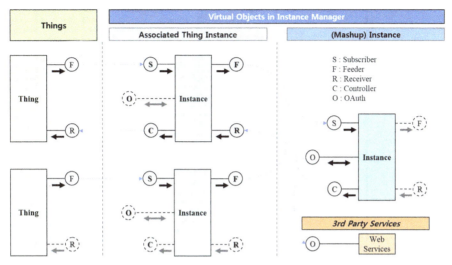

Figure 1   Overall interfaces for instances hosting.

Thirdly, 'Controller' interface' provides a function to deliver messages to outside directly. It is used to deliver message which has information to other instance or send a message about controlling order to certain IoT device.

Lastly, 'OAuth' interface provides a function of connecting outside 3rd Party service. Thus, it is used to send and receive a message through 3rd Party service.

The reasons of presenting call out structure of indicative interface to instance manager are for designing interface more simply in the view of virtual object developer, for using call out between instances and for blocking mal-function in advance through controlling message call out between instances in the view of instance manager.

3.3 Instance Manager

Instance manager provides a function to execute logic of virtual object in a form of each JavaScript instance. Instance manager consists of main modules like Figure 2 and the main module related to create and run instance based on virtualized object consists of Interface Server (providing a function of communicating between instance manager and outside modules), Message Router (providing a function of routing read and write message between instances), Instance Lifecycle Manager (providing a function of controlling

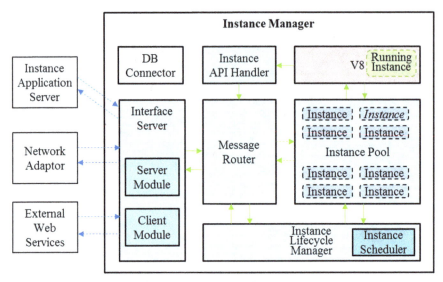

Figure 2   Architecture of the instance manager.

and scheduling instance lifecycle), Instance Pool (providing a function of pooling contexts of several instances), V8 (providing a function of executing instance).

Another module, IAS, provides web service for controlling instance manager in a web-based form and deploying function setting up a virtual object file. Network Adapter is responsible for managing messages between receiving device through HTTP, ZigBee interface and External Web Services is responsible for connection of 3rd Party web service.

Instance manager adjusts simultaneously handling quantity of instance for executing several virtualized objects effectively in an environment of embedded system and it should have the structure of providing quick response about each instance execution. It can be possible by pooling context information about the instance using instance pool after the first instance execution and by minimizing context creation time when instance is called out.

Since avoiding additional I/O usage except for the case of call out instance at first, performance can be higher.

These structure especially in IoT environment, is suitable to a service pattern which has a lot of message reception and handling between devices frequently. Therefore, it can be possible to use pooled context information to instances which contains frequent call out periodically and minimize creation time of the context by binding JavaScript engine directly.

3.4 Implementation and Performance

To evaluate the performance of the suggested Instance Hosting platform, we used an ASUS RT-N61 Wi-Fi AP which does support the customizable embedded OS, and added a TI ZigBee adapter to the internal serial port to support legacy non-IoT gadgets.

After porting the GNU/Linux based DebWRT embedded OS to the Wi-Fi AP, we built the Instance Manager using C++ language and the Instance Application Server using Apache-Rewrite Module, PHP5, MySQL and Python, and installed them on the AP system. The VO execution handling module was developed using V8 engine.

The execution logic for VO is described in JavaScript language and is processed on the V8 JavaScript engine. At the first time when a VO is to be executed, the actual object instance is created according to the VO description and this object instance is handled in the Instance Pool and used for the specific IoT gadget virtualization.

To evaluate the performance of the Instance Hosting platform, we conducted some experiments in real working environment, and some of the experiment results are shown in Figures 3 and 4. (During the experiment, each IoT gadget changes its state in every 10 seconds, i.e. its instance is invoked in every 10 seconds.)

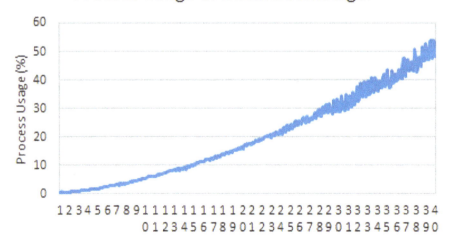

Figure 3 Memory consumption by the deployed number of instance increases.

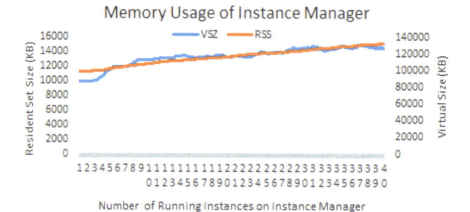

Figure 4   CPU consumption by the deployed number of instance increases.

As the number of hosted instance increases, the process usage (shown in Figure 3) and the memory usage (shown in Figure 4) are generally increased, but keep the stable behaviors. The initial memory consumption to execute the Instance Hosting platform is about 11 MB, but additional memory consumption for each instance is only less than 100 KB. This means that our proposed Instance Hosting platform can handle the enough number of IoT gadgets in home environments and the extended number of mashup instance under the limited embedded system environment.

4 Edge Cloud and SDN/NFV

So far, how to create and control the VOs in light-weight embedded system is recognized and it is confirmed that these structure works as expected. However, there can be a limitation to guarantee this performance continuously in light-weight embedded system at the situation of increasing IoT devices continuously at user side. Especially, the longer time user uses the IoT devices, the earlier storage which can afford the accumulation of data of intelligent service and previous data history for advanced intelligent service and process capacity approaches to the limit.

If Instance Hosting platform is implemented in SDN/NFV, a structure of Instance Hosting platform, as suggested before, could be changed as shown in Figure 5.

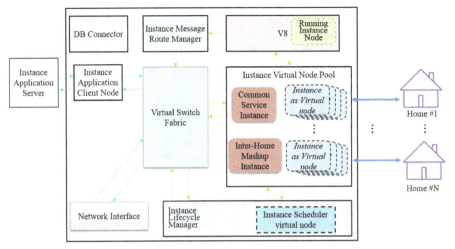

Figure 5 Instance hosting platform architecture over SDN/NFV platform.

Therefore, it is necessary that Edge Cloud which can provide intelligent mashup service storing and controlling data from several IoT devices and existing at the edge of network. But, it is not a realistic approach to set up the equipment for new Edge Cloud just for the connecting service between IoT devices at the edge of network. Therefore, loading Instance Hosting platform for connecting the IoT devices in approach of SDN/NFV framework which comes into the spotlight these days.

Interface Server for affording each IoT devices in Instance Hosting platform can be replaced to Network interface function module of SDN/NFV. Message Router in an instance manager can apply efficient router function module implementing at SDN/NFV when each VO Instances and Scheduler instances make to become virtual network entity in SDN/NFV. Instance connecting common outside service (virtualized instance which has information of local weather, regional traffic) which should be existed for each instance manager can be made to become general VO which can be used in common between several instance managers.

In addition, Point of Exist (POE) which has previous instance hosting platform existed on the physical edge of outside network and each user network on virtual edge of private/public network on the structure of SDN/NFV network exists on the virtual edge of private/public network in structure of SDN/NFV network when instance hosting platform over SDN/NFV platform is constructed. Therefore, there are additional advantages about connection which can be new between many instances hosting manager. (It becomes to edge point since virtual edges of each users' network-public network co-exist physically in SDN/NFV platform finally. Therefore, mashup service which can cover all network of various each user can be possible at this edge point.)

5 Conclusion

IoT gadgets especially for individual person and smart home might make our daily lives more convenient and efficient. But, the increased number of personal belongings which needs to be managed also induces another kind of burdens. In this paper, we suggest the IoT management platform that can be located in the light-weight embedded system (like a Wi-Fi AP system) and handles the configuration, control and mashup service creation for several IoT VOs with user friend Web technologies. And by the experiments in real environment, we show that the suggested platform can provide the designed services with stable performance. In near future, this kind of IoT management platform can be deployed on Edge Clouds and SDN/NFV infrastructure.

Acknowledgment

This work was supported by Institute for Information & communications Technology Promotion (IITP) grant funded by the Korea government (MSIP) (No. B0101-16-0331, Development of Multi-Screen Adaptive Autonomous Smart Media Technology).

References

[1] *"Appcessory Economics: Enabling loosely coupled hardware/software innovation,"* Cornell University Library, Oct. 28, 2013, from http://arxiv.org/abs/1209.5901

[2] Andreas Nettsträter, *"Deliverable D1.3-Updated Reference model for IoT v1.5,"* IoT-A, Jul. 16, 2012

[3] Rob Chandhok, *"A fast track to the Internet of Everything,"* Oct. 18, 2013 from https://www.alljoyn.org/sites/default/files/alljoyn-alliance.pdf

[4] Xively, *"Public Cloud for Internet of Things,"* https://xively.com/

[5] IFTTT, *"IF This Than That,"* http://www.ifttt.com

[6] S. Duquennoy, G. Grimaud, and J.-J. Vandewalle, *"The Web of Things: Interconnecting Devices with High Usability and Performance."* In Proc. ICESS, HangZhou, Zhejiang, China, May 2009.

[7] D. Guinard, V. Trifa, and E. Wilde. *"A Resource Oriented Architecture for the Web of Things."* In Proc. IoT, Tokyo, Japan, 2010.

[8] J. Park and N. Kang, *"Entity authentication scheme for secure WEB of Things applications,"* J. KICS, vol. 38B, no. 5, pp. 394–400, May 2013.

[9] P. Levis and D. Culler, *"Mate: A Tiny Virtual Machine for Sensor Networks,"* In Proceedings of the International Conference on Architectural Support for Programming Languages and Operating Systems, ASPLOS-X, October, 2002.

[10] N. Brouwers, K. Langendoen, and P. Corke. *"Darjeeling, a feature-rich VM for the resource poor,"* In Proc. SenSys, Berkeley, California, 2009.

[11] Youngjun Kim, Yongkeun Jeon, and Ilyoung Chong, *"Device Objectification and Orchestration Mechanism for IoT Intelligent Service,"* J. KICS, vol. 38C, no. 1, pp. 19–32, January 2013.

[12] Kovatsch, M, Lanter, M., Duquennoy, S., *"Actinium: A RESTful runtime container for scriptable Internet of Things applications,"* Internet of Things (IOT), 2012 3rd International Conference on the, pp. 135–142, Oct. 24–26, 2012.

[13] Guinard, D., Trifa, V., Wilde, E., *"A Resource Oriented Architecture for the Web of Things,"* Internet of Things (IoT) 2010, pp. 1–8, Nov. 29, 2010.

[14] V. Trifa, S. Wieland and D. Guinard, *"Towards the Web of Things: Web Mashups for Embedded Devices,"* Second Workshop on Mashups, Enterprise Mashups and Lightweight Composition on the Web, Madrid, Apr. 20–24, 2009.

Biographies

J. Yang (S'05) received M.S. in computer science from InJe University in 2005 and HERIT Inc. in 2008 and currently he is Ph.D. candidate student in Korea Advanced Institute of Science and Technology (KAIST). His main research interests include next generation network, multimedia streaming issues, and IoT.

H. Park (S'07) received M.S. in communications engineering from Information and Communications University in 2007 and currently, she is Ph.D. candidate student in Korea Advanced Institute of Science and Technology (KAIST). Her main research interests include IPTV, broadcast networks, multimedia streaming issues, and next generation network.

S. K. Jung received Ph.D. in electrical engineering from Korea Advanced Institute of Science and Technology (KAIST) in 2007.

Currently, He is team leader in KAIST KI ITC. His main research interests include Internet of Things, Smart Devices and next generation network.

H. Kim received his Ph.D. degree in the department of information engineering from Sungkyunkwan University, Korea, in 2005. From 1992 to 2002, he was senior engineer in Electronics and Telecommunications Research Institute (ETRI), Daejon, Korea. He has been a professor in the department of computer science in Namseoul University, Korea. His research interests include cloud computing, next generation network and embedded computing systems.

Analysis of Dyscalculia Evidences through Artificial Intelligence Systems

F. Ferraz[1], H. Vicente[2,3], A. Costa[2] and J. Neves[2]

[1]Departamento de Informática, Universidade do Minho, Braga, Portugal
[2]Centro Algoritmi, Universidade do Minho, Braga, Portugal
[3]Departamento de Química, Escola de Ciências e Tecnologia,
Universidade de Évora, Évora, Portugal
E-mail: filipatferraz@gmail.com; hvicente@uevora.pt;
{acosta; jneves}@di.uminho.pt

Received 5 August 2016; Accepted 17 October 2016;
Published 4 November 2016

Abstract

Dyscalculia is usually perceived of as a specific learning difficulty for mathematics or, more appropriately, arithmetic. Because definitions and diagnoses of dyscalculia are in their infancy and sometimes are contradictory. However, mathematical learning difficulties are certainly not in their infancy and are very prevalent and often devastating in their impact. Co-occurrence of learning disorders appears to be the rule rather than the exception. Co-occurrence is generally assumed to be a consequence of risk factors that are shared between disorders, for example, working memory. However, it should not be assumed that all dyslexics have problems with mathematics, although the percentage may be very high, or that all dyscalculics have problems with reading and writing. Because mathematics is very developmental, any insecurity or uncertainty in early topics will impact on later topics, hence to need to take intervention back to basics. However, it may be worked out in order to decrease its degree of severity. For example, *disMAT*, an *app* developed for *android* may help children to apply mathematical concepts, without much effort, that is turning in itself, a promising tool to dyscalculia treatment. Thus, this work will focus on the development of a *Decision Support*

Journal of Software Networking, 53–78.
doi: 10.13052/jsn2445-9739.2016.004

System to estimate children evidences of dyscalculia, based on data obtained on-the-fly with *disMAT*. The computational framework is built on top of a *Logic Programming* approach to *Knowledge Representation and Reasoning*, grounded on a *Case-based* approach to computing, that allows for the handling of incomplete, unknown, or even self-contradictory information.

Keywords: Dyscalculia, Logic Programming, Knowledge Representation and Reasoning, Case-based Computing, Decision Support Systems.

1 Introduction

Dyscalculia was primarily defined as "a structural disorder of mathematical abilities" [1]. Therefore, and through some consequent studies, it can be defined as a mathematical learning disability that affects the ability to perform operations and make the proper use of arithmetic. Frequently described as "the dyslexia or blindness for numbers", the dyscalculia is hard to be well-diagnosed, despite the incidence on 6 to 7% in the population [2]. Nevertheless, it is relevant to distinguish the type of dyscalculia under analysis:

- The person in question suffered a trauma, like an injury or a stoke, and developed this difficulty in leading with numbers, then it is called *acalculia*; and
- The disorder exists since birth, with the absence of accidents, then it is called *developmental dyscalculia*, since it will accompanied the individual through ages. This last type it is the most common and the one that is referred to as *dyscalculia* [3].

Besides affecting the realization of simple calculations with two digits and basic operations (i.e., addition, subtraction, multiplication and division), dyscalculia also influences tasks such as to distinguish left from right, to express the time, or even to count money/cash. Since this specific developmental disorder can be reflected in various areas of mathematic, dyscalculia may be set in six sub-areas, taking into account the most affected ones [1], namely:

- *Lexical dyscalculia* – troubles in reading mathematical symbols;
- *Verbal dyscalculia* – troubles in naming mathematical quantities, numbers and symbols;
- *Graphic dyscalculia* – troubles in writing mathematical symbols;
- *Operational dyscalculia* – troubles in performing mathematical operations and calculus;

- *Practognostic dyscalculia* – troubles in enumerating, manipulating and comparing real objects and pictures; and
- *Ideagnostic dyscalculia* – troubles in mental operations and in the understanding of mathematical concepts.

Regarding origins, some studies claims that the intraparietal sulcus is the area accountable for the number sense, which means that the dyscalculics present a poor activity in this brain area [4]. Conversely, geneticists propose a theory that there is a gene responsible for the transmission of this disorder through heritage, although this hypothesis is not properly proved. So, after the screening tests for dyslexia, the psychologist suggests to consult a neurologist to perform an fMRI in order to correctly diagnose if dyscalculia it is the case [5].

Additionally, dyscalculia may also be classified according to the state of neurological immaturity [6, 7], namely:

- As a *former state*, related with the individuals that react favorably to therapeutically intervention;
- As a *second one*, associated to the individuals who have other learning disabilities; and
- As a *last one*, linked to the individuals that feature an intellectual deficit caused by a neurological injury(ies).

Dyscalculia is irreversible, i.e., the disorder cannot be treated, but can be worked out in order to decrease its degree of severity. This involves the methodic habit to solve exercises regarding dyscalculia's issues, like exercising memory, counting amounts, and so one, leading the subject to evolve his/her weaknesses [8].

Since dyscalculia can affect daily life, it is recommended to screen this disability in early stages of life. Afterthought, the treatment or therapeutics must commence as soon as possible, in order to provide a regular routine to dyscalculic. Therefore, the therapeutics must use attractive methods to help the children to deal with the mathematical issues [9]. In order to meet this challenge, an *android app*, *disMAT*, was developed to improve the arithmetic skills of children from the age range 5–10 through simple tasks that look at their weak spots, like measures, without much effort, obligation nor awareness, turning this support system into a promising tool [10]. *disMAT* is a simple game, where the child has to choose the level according to its difficulty and then solve every one of the nine level's tasks, trying to hit the maximum score. It presents a puzzle at the end of each level to confront the child – it is a bonus task. Besides helping the child evolve the mental calculus, this *app* is also attractive with respect to the fact that children nowadays are keeping up with the technologic era, carrying their tablets and smartphones everywhere to entertainment.

Indeed, this paper addresses the theme of dyscalculia and describes an attempt to diagnosis the disorder using a *Case-based* (*CB*) approach to computing [11, 12]. The *app disMAT* was applied to a group of children within the required age range, and some parameters were recorded aiming to build up a knowledge base. To set the structure of the environment and the associate inference mechanisms, a computational framework centred on a *Logic Programming* (*LP*) based approach to knowledge representation and reasoning was used [13]. It may handle unknown, incomplete, or even contradictory data, information or knowledge.

2 Knowledge Representation

Many approaches to knowledge representation have been proposed using the *Logic Programming* (*LP*) epitome, namely in the area of *Model Theory* [14, 15], and *Proof Theory* [13, 16]. In the present work the *Proof Theoretical* approach in terms of an extension to the *LP* language is followed, where a logic program is a finite set of clauses, given in the form:

$$\{$$

$$\neg\, p \leftarrow not\ p,\ not\ exception_p$$

$$p \leftarrow p_1, ..., p_n, not\ q_1, ..., not\ q_m$$

$$?\,(p_1, ..., p_n, not\ q_1, ..., not\ q_m)\ \ (n,\ m \geq 0)$$

$$exception_{p_1}$$

$$...$$

$$exception_{p_j}\ \ (0 \leq j \leq k),\ \ being\ k\ and\ integer\ number$$

$$\} :: scoring_{value}$$

where the first clause stand for predicate's closure, "," denotes "*logical and*", while "*?*" is a domain atom denoting falsity. The p_i, q_j, and p are classical ground literals, i.e., either positive atoms or atoms preceded by the classical negation sign $\neg$ [13]. Indeed, $\neg$ stands for a strong declaration that speaks for itself, and *not* denotes *negation-by-failure*, or in other words, a flop in proving a given statement, once it was not declared explicitly. Under symbols' theory, every program is associated with a set of abducibles [14, 15], given here in the form of exceptions to the extensions of the predicates that make the program, i.e., clauses of the form:

$$exception_{p_1}, ..., exception_{p_j} \ (0 \leq j \leq k), \ being \ k \ an \ integer \ number$$

that stand for data, information or knowledge that cannot be ruled out. On the other hand, clauses of the type:

$$? \ (p_1, ..., p_n, \ not \ q_1, ..., not \ q_m) \ \ (n, \ m \geq 0)$$

also named invariants or restrictions, allows one to set the context under which the universe of discourse has to be understood. The term $scoring_{value}$ stands for the relative weight of the extension of a specific *predicate* with respect to the extensions of peers ones that make the inclusive or global program.

2.1 Quantitative Knowledge

In order to set one's approach to knowledge representation, two metrics will be set, namely the Quality-of-Information (*QoI*) of a logic program that will be understood as a mathematical function that will return a truth-value ranging between 0 and 1 [17, 18], once it is fed with the extension of a given predicate, i.e., $QoI_i = 1$ when the information is *known* (*positive*) or *false* (*negative*) and $QoI_i = 0$ if the information is *unknown*. For situations where the extensions of the predicates that make the program also include *abducible* sets, its terms (or clauses) present a $QoI_i \epsilon [0, 1]$, in the form:

$$QoI_i = {}^1/Card \tag{1}$$

if the *abducible* set for *predicates* i and j satisfy the *invariant*:

$$? \left(\left(exception_{p_i}; \ exception_{p_j} \right), \neg \left(exception_{p_i}; \ exception_{p_j} \right) \right)$$

where ";" denotes "*logical or*" and "*Card*" stands for set cardinality, being $i \neq j$ and $i, j \geq 1$ (a pictorial view of this process is given in Figure 1(a), as a pie chart).

On the other hand, the clauses cardinality (K) will be given by $C_1^{Card} + ... + C_{Card}^{Card}$, if there is no constraint on the possible combinations among the abducible clauses, being the QoI acknowledged as:

$$QoI_{i_{1 \leq i \leq Card}} = {}^1/C_1^{Card}, ..., {}^1/C_{Card}^{Card} \tag{2}$$

where C_{Card}^{Card} is a card-combination subset, with *Card* elements. A pictorial view of this process is given in Figure 1(b), as a pie chart.

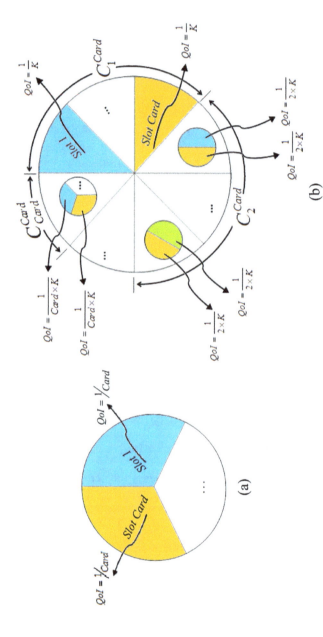

Figure 1 *QoI's* values for the abducible set for *predicate*$_i$ with (a) and without (b) constraints on the possible combinations among the abducible clauses.

However, a term's *QoI* also depends on their attribute's *QoI*. In order to evaluate this metric, look to Figure 2, where the segment with bounds 0 and 1 stands for every attribute domain, i.e., all the attributes range in the interval [0, 1]. [*A, B*] denotes the range where the unknown attributes values for a given predicate may occur (Figure 2). Therefore, the *QoI* of each attribute's clause is calculated using:

$$QoI_{attribute_i} = 1 - \|A - B\| \tag{3}$$

where $\|A{-}B\|$ stands for the modulus of the arithmetic difference between *A* and *B*. Thus, in Figure 3 is showed the *QoI's* values for the abducible set for *predicate_i*.

Under this setting, another metric has to be considered, which will be denoted as *DoC* (*Degree-of-Confidence*), that stands for one's confidence that the argument values or attributes of the terms that make the extension of a given predicate, having into consideration their domains, are in a given interval [19]. The *DoC* is figured using $DoC = \sqrt{1 - \Delta l^2}$, where Δl stands for the argument interval length, which was set to the interval [0, 1] (Figure 4).

Thus, the universe of discourse is engendered according to the information presented in the extensions of such predicates, according to productions of the type:

$$predicate_i - \bigcup_{1 \leq j \leq m} clause_j \left(\left((A_{x_1}, B_{x_1}) \left(QoI_{x_1}, DoC_{x_1} \right) \right), ..., \right.$$
$$\left. \left((A_{x_l}, B_{x_l}) \left(QoI_{x_l}, DoC_{x_l} \right) \right) \right) :: QoI_j :: DoC_j \tag{4}$$

where $\bigcup$, *m* and *l* stand, respectively, for *set union*, the *cardinality* of the extension of *predicate_i* and the number of attributes of each clause [19]. On the other hand, either the subscripts of the QoI_s and the DoC_s, or those of the pairs (A_s, B_s), i.e., $x_1, ..., x_l$, stand for the attributes' clauses values ranges.

2.2 Qualitative Knowledge

In present study both qualitative and quantitative data/knowledge are present. Aiming at the quantification of the qualitative part and in order to make easy the understanding of the process, it will be presented in a graphical form. Taking as an example a set of *n* issues regarding a particular subject (where there are *k*

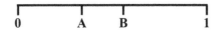

Figure 2   Setting the *QoIs* of each attribute's clause.

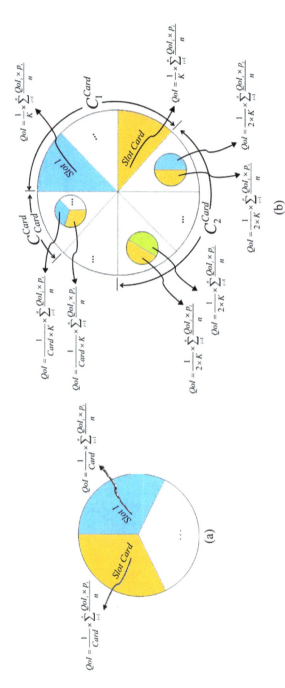

Figure 3 *QoI's* values for the abducible set for *predicate_i* with (a) and without (b) constraints on the possible combinations among the abducible clauses. $\sum_{i=1}^{n}(QoI_i \times p_i)/n$ denotes the *QoI's* average of the attributes of each clause (or term) that sets the extension of the predicate under analysis. n and p_i stand for, respectively, for the attribute's cardinality and the relative weight of attribute p_i with respect to its peers ($\sum_{i=1}^{n} p_i = 1$).

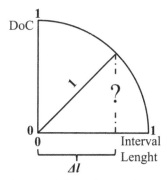

Figure 4 Evaluation of the attributes' degree of confidence.

possible choices, i.e., *none*, *low*, ..., *high* and *very high*), let us itemized an unitary area circle split into n slices (Figure 5). The marks in the axis correspond to each of the possible options. If the answer to issue 1 is *high* the correspondent area is $\pi \times \left(\sqrt{\frac{k-1}{k \times \pi}}\right)^2 /n$, i.e., $(k-1)/(k \times n)$ (Figure 5(a)). Assuming that in the issue 2 are chosen the alternatives *high* and *very high*, the correspondent area ranges between $\left[\pi \times \left(\sqrt{\frac{k-1}{k \times \pi}}\right)^2 /n, \pi \times \left(\sqrt{\frac{k}{k \times \pi}}\right)^2 /n\right]$, i.e., $[(k-1)/(k \times n), \ k/(k \times n)]$ (Figure 5(b)). Finally, in issue n if no alternative is ticked, all the hypotheses should be considered and the area varies in the interval $\left[0, \ \pi \times \left(\sqrt{\frac{k}{k \times \pi}}\right)^2 /n\right]$, i.e., $[0, \ k/(k \times n)]$ (Figure 5(c)). Thus, the total area is the sum of the partial ones (Figure 5(d)), i.e., $[(2k-2)/(k \times n), \ (3k-1)/(k \times n)]$.

3 Case-based Computing

The *Case-Based (CB)* approach to computing stands for an act of finding and justifying a solution to a given problem based on the consideration of the solutions of similar past ones, either using old solutions, or by reprocessing and generating new data or knowledge from the old ones [11, 12]. In *CB* the *cases* are stored in a *Case-base*, and those cases that are similar (or close) to a new one are used in the problem solving process.

The typical *CB* cycle (Figure 6) presents the mechanism that should be followed to have a consistent model. The first stage entails an initial description and a reprocessing of the problem's data or knowledge. The new case is defined

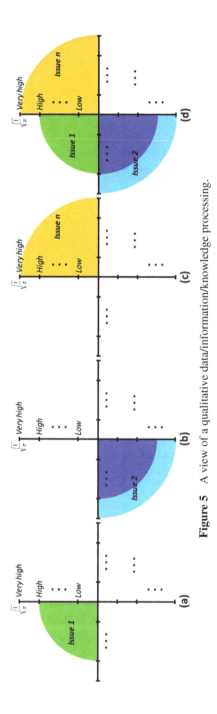

Figure 5   A view of a qualitative data/information/knowledge processing.

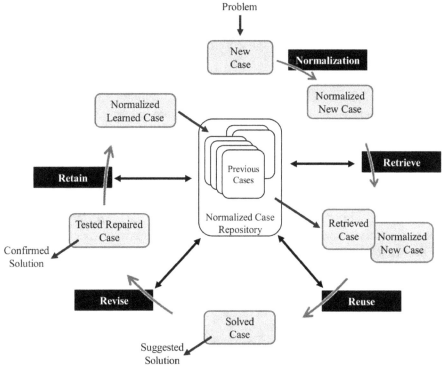

Figure 6 The typical *CB* cycle (adapted from [11]).

and it is used to retrieve one or more cases from the repository. At this point it is important to identify the characteristics of the new problem and retrieve cases with a higher degree of similarity to it. Thereafter, a solution to the problem emerges, on the *Reuse* phase, based on the blend of the new case with the retrieved ones. The suggested solution is reused (i.e., adapted to the new case), and a solution is provided [11, 12]. However, when adapting the solution it is crucial to have feedback from the user, since automatic adaptation in existing systems is almost impossible. This is the *Revise* stage, in which the suggested solution is tested by the user, allowing for its correction, adaptation and/or modification, originating the test repaired case that sets the solution to the new problem. The test repaired case must be correctly tested to ensure that the solution is indeed correct. Thus, one is faced with an iterative process since the solution must be tested and adapted, while the result of considering that solution is inconclusive. During the *Retain* (or *Learning*) stage the case is learned and the knowledge base is updated with the new case [11, 12].

Undeniably, despite promising results, the current *CB* systems are neither complete nor adaptable enough for all domains. In some cases, the user cannot choose the similarity(ies) method(s) used in the retrieval phase and is required to follow the system defined one(s), even if they do not meet their needs. Moreover, in real problems, the access to all necessary information is not always possible, since existent *CB* systems have limitations related to the capability of dealing, explicitly, with unknown, incomplete, and even self-contradictory information. To make a change, a different *CB* cycle was induced (Figure 7). It takes into consideration the case's *QoI* and *DoC* metrics. It also contemplates a cases optimization process present in the *Case-base*, whenever they do not comply with the terms under which a given problem as to be addressed (e.g., the expected *DoC* on a prediction was not attained). This process that uses either *Artificial Neural Networks* [20, 21], *Particle Swarm Optimization* [22] or *Genetic Algorithms* [16], just to name a few, generates a set of new cases which must be in conformity with the invariant:

$$\bigcap_{i=1}^{n} (\boldsymbol{B_i}, \boldsymbol{E_i}) \neq \emptyset \tag{5}$$

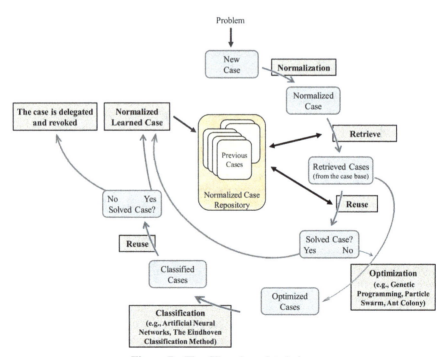

Figure 7   The *CB* cycle updated view.

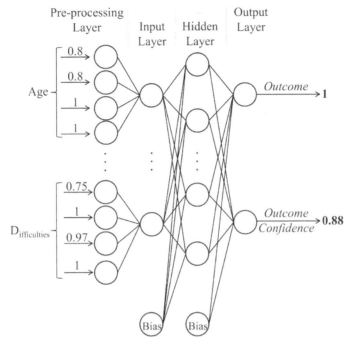

Figure 8 A case's classification procedure based on *ANNs*.

i.e., it states that the intersection of the attribute's values ranges for the cases' set that make the *Case-base* or their optimized counterparts (B_i) (being *n* its cardinality), and the ones that were object of a process of optimization (E_i), cannot be empty (Figure 7). In this *CB* cycle are used *Artificial Neural Networks* (*ANNs*) for case's classification in the following way:

- The extremes of the attribute's values ranges, as well as their *DoCs* and *QoIs* are fed to the *ANN*; and
- The outputs are given in a form that ensures that the case may be used to solve the problem (*no* (*0*), *yes* (*1*)), and a measure of the system confidence on such a result is provided (Figure 8).

4 Methods

4.1 Participants

The data was taken from the evaluation of 203 children of primary schools in the North of Portugal who played the *disMAT app* [10]. The children enrolled in this study aged between 5 to 10 years old, with an average of 8.2 years old. The gender distribution was 42.4% and 57.6% for male and female, respectively. Forty eight students, i.e., 23.6% of the cohort, were signalized as having difficulties with numbers and mathematical concepts.

Students participated in the study voluntarily without any pressure or coercion and were informed that their grades would not be affected. Each of the participants gave an informed consent form to participate in the study. The study was conducted in compliance with the institutional guidelines For each participant was recorded the age, the number of game levels completed, the minimum score obtained, as well as the maximum score, the response time in each of the three levels and the classification of understanding and doing difficulties through the game.

4.2 A Logic Programming Approach to Data Processing

The knowledge database is given in terms of the extensions of the tables depicted in Figure 9, which stand for a situation where one has to manage information about children evidences of dyscalculia. Under this scenario some incomplete and/or unknown data is also present. For instance, the *Level 3 Response Time* in case 1 is unknown, which is depicted by the symbol $\perp$, while the opinion about *Understanding Difficulties* is not conclusive (*Very Easy/Easy*).

Applying the algorithm presented in [19] to the table or relation's fields that make the knowledge base for dyscalculia diagnosis (Figure 9), and looking to the *DoCs* values obtained as described before, it is possible to set the arguments of the predicate *diagnostic* (*diag*) referred to below, whose extensions also denote the objective function with respect to the problem under analyze:

$$diag: Age, \; Levels Completed, \; Min_{imum \; Score}, \; Ave_{rage \; Score},$$

$$Max_{imum \; Score}, Level1 \; R_{esponse \; Time}, Level2 \; R_{esponse \; Time},$$

$$Level3 \; R_{esponse \; Time}, D_{ifficulties} \; \rightarrow \; \{0, 1\}$$

The algorithm presented in [19] encompasses different phases. In the former one the clauses or terms that make extension of the predicate under study are established. In a second step the boundaries of the attributes intervals

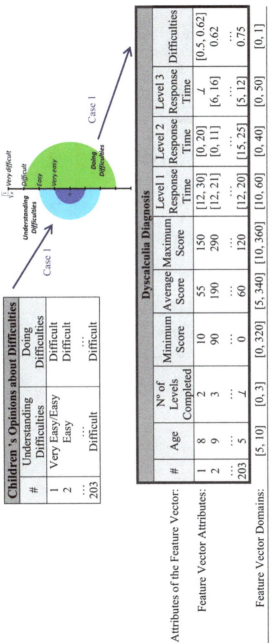

Figure 9 A fragment of the knowledge base for dyscalculia diagnosis.

are set in the interval $[0, 1]$ according to a normalization process given by the expression $(Y - Y_{min})/(Y_{max} - Y_{min})$, where the Y_s stand for themselves. Finally, the *DoC* is evaluated as described in Section 2.1.

Exemplifying the application of the algorithm presented in [19], to a term (a *disMAT app user*) that presents the feature vector ($Age = 9$, $Levels Completed = 1$, $Minimum\ Score = 5$, $Average\ Score = 45$, $Maximum\ Score = 60$, $Level 1 Response\ Time = [18, 30]$, $Level 2\ Response\ Time = \bot$, $Level 3\ Response\ Time = 0$, $Difficulties = [0.75, 1]$), one may have:

Begin (DoCs evaluation)
The predicate's extension that sets the Universe-of-Discourse to the case (term) under observation is fixed

{

$\neg\ diag\ \left(\left(\left(A_{Age},\ B_{Age}\right)\left(QoI_{Age},\ DoC_{Age}\right)\right), ..., \left(\left(A_D,\ B_D\right)\left(QoI_D,\ DoC_D\right)\right)\right)$

$\leftarrow not\ diag\ \left(\left(\left(A_{Age},\ B_{Age}\right)\left(QoI_{Age},\ DoC_{Age}\right)\right), ..., \left(\left(A_D,\ B_D\right)\left(QoI_D,\ DoC_D\right)\right)\right)$

$diag\ \left(\left(\left(9,\ 9\right)\left(1_{[9,\ 9]},\ DoC_{[9,\ 9]}\right)\right), ..., \left(\left(0.75,\ 1\right)\left(QoI_{[0.75,\ 1]},\ DoC_{[0.75,\ 0.1]}\right)\right)\right)$

$:: 1 :: DoC$

$\underbrace{[5, 10] \qquad\qquad\qquad \cdots \qquad [0,\ 1]}_{attribute's\ domains}$

} :: 1

The attribute's boundaries are set to the interval $[0, 1]$, according to a normalization process that uses the expression $(Y - Y_{min})/(Y_{max} - Y_{min})$

{

$\neg diag\ \left(\left(\left(A_{Age},\ B_{Age}\right)\left(QoI_{Age},\ DoC_{Age}\right)\right), ..., \left(\left(A_D,\ B_D\right)\left(QoI_D,\ DoC_D\right)\right)\right)$

$\leftarrow not\ diag\ \left(\left(\left(A_{Age},\ B_{Age}\right)\left(QoI_{Age},\ DoC_{Age}\right)\right), ..., \left(\left(A_D,\ B_D\right)\left(QoI_D,\ DoC_D\right)\right)\right)$

$diag\ \left(\left(\left(0.8,\ 0.8\right)\left(1_{[0.8,\ 0.8]}, DoC_{[0.8,\ 0.8]}\right)\right), ..., \left(\left(0.75,\ 1\right)\left(1_{[0.75,\ 1]},\ DoC_{[0.75,\ 1]}\right)\right)\right)$

$:: 1 :: DoC$

$\underbrace{[0,\ 1] \qquad\qquad\qquad ... \qquad [0,\ 1]}_{attribute's\ domains\ once\ normalized}$

} :: 1

The DoC's values are evaluated

{

$\neg\, diag\ \left(\left(\left(A_{Age},\ B_{Age}\right)\left(QoI_{Age},\ DoC_{Age}\right)\right),...,\left(\left(A_D,\ B_D\right)\left(QoI_D,\ DoC_D\right)\right)\right)$

$\leftarrow not\ diag\ \left(\left(\left(A_{Age},\ B_{Age}\right)\left(QoI_{Age},\ DoC_{Age}\right)\right),...,\left(\left(A_D,\ B_D\right)\left(QoI_D,\ DoC_D\right)\right)\right)$

$diag\ \underbrace{\left(\left(\left(0.8,\ 0.8\right)\left(1,\ 1\right)\right),...,\left(\left(0.75,\ 1\right)\left(1,\ 0.97\right)\right)\right)}\ ::\ 1\ ::\ 0.88$

attribute's values ranges once normalized and
respective QoI and DoC values

$\underbrace{\begin{matrix}[0,\ 1] & \cdots & [0,\ 1]\end{matrix}}$
attribute's domains once normalized

} :: 1

End

5 Results and Discussion

5.1 A Case-based Approach to Computing

In this section is set the model of the universe of discourse, where the computational part is based on a *CB* approach to computing. Contrasting with other problem solving tools (e.g., those that use *Decision Trees* or *Artificial Neural Networks*), relatively little work is done offline [23]. Undeniably, in almost all the situations the work is performed at query time. The main difference between this approach and the typical *CB* one relies on the fact that not only all the cases have their arguments set in the interval [0, 1], a situation that is complemented with the prospect of handling incomplete, unknown, or even self-contradictory data, information or knowledge. Thus, the classic *CB* cycle was changed (Figure 7), being the *Case-base* given in terms of the pattern:

$$Case = \{< Raw_{data},\ Normalized_{data},\ Description_{data} >\}$$

where the $Description_{data}$ field will not be object of attention in this study. Undeniably, when confronted with a new case, the system is able to retrieve all cases that meet such a case structure and optimize, when necessary, such a population, i.e., it considers the attributes *DoC*'s value of each case or of their optimized counterparts when analyzing similarities among them.

Thus, under the occurrence of a new case, the goal is to find similar cases in the *Case-base*. Having this in mind, the algorithm described above is applied to a new case, that presents the feature vector ($Age = 8$, $L_{evels}C_{ompleted} = 3$, $Min_{imum\ Score} = \perp$, $Ave_{rage\ Score} = 150$, $Max_{imum\ Score} = 330$, $L_{evel}1$ $R_{esponse\ Time} = 15$, $L_{evel}2\ R_{esponse\ Time} = [6,\ 20]$, $L_{evel}3\ R_{esponse\ Time} = [15, 35]$, $D_{ifficulties} = [0.88, 1]$), having in consideration that the cases retrieved from the *Case-base* satisfy the invariant presented in Equation (5), in order to ensure that the intersection of the attributes range in the cases that make the *Case-base* repository or their optimized counterparts, and the equals in the new case cannot be empty. Then, the computational process may be continued, with the outcome:

$$diag_{new} \underbrace{(((0.6,\ 0.6)\ (1,\ 1)),\ ...,\ ((0.88,\ 1)\ (1,\ 0.99)))}_{new\ case} :: 1 :: 0.87$$

Now, the *new case* may be portrayed on the *Cartesian* plane in terms of its *QoI* and *DoC*, and by using clustering methods [24] it is feasible to identify the cluster(s) that intermingle with the *new one* (epitomized as a square in Figure 10). The *new case* is compared with every retrieved case from the clusters using a similarity function *sim*, given in terms of the average of the

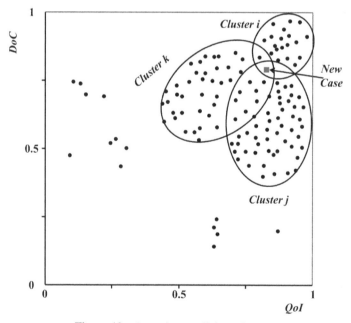

Figure 10   A case's set split into clusters.

modulus of the arithmetic difference between the arguments of each case of the selected cluster and those of the *new case*. Thus, one may have:

$$diag_1 \left(\left(\left(0.8, \ 0.8\right)\left(1, \ 1\right)\right), ..., \left(\left(0.75, \ 0.88\right)\left(1, \ 0.99\right)\right)\right) :: 1 :: 0.92$$
$$diag_2 \left(\left(\left(0.6, \ 0.6\right)\left(1, \ 1\right)\right), ..., \left(\left(0.25, \ 0.75\right)\left(1, \ 0.87\right)\right)\right) :: 1 :: 0.84$$
$$\vdots$$
$$\underbrace{diag_j \left(\left(\left(0.8, \ 0.8\right)\left(1, \ 1\right)\right), ..., \left(\left(0.5, \ 0.5\right)\left(1, \ 1\right)\right)\right) :: 1 :: 0.85}_{normalized\ cases\ from\ retrieved\ cluster}$$

Assuming that every attribute has equal weight, for the sake of presentation, the dissimilarity between $diag_{new}$ and the $diag_1$, i.e., $diag_{new \longrightarrow 1}$, may be computed as follows:

$$diag_{new \to 1}^{DoC} = \frac{\|1 - 1\| + ... + \|0.99 - 0.99\|}{9} = 0.11$$

Thus, the similarity for $diag_{new \to 1}^{DoC}$ is set as *1 – 0.11 = 0.89*. Regarding *QoI* the procedure is similar, returning $diag_{new \to 1}^{QoI} = 1$. Thus, one may have:

$$diag_{new \to 1}^{QoI, DoC} = \frac{1 + 0.89}{2} = 0.94$$

These procedures should be applied to the remaining cases of the retrieved clusters in order to obtain the most similar ones, which may stand for the possible solutions to the problem.

5.2 Model Validation and Performance Assessment

A common tool to evaluate the results presented by the classification models is the coincidence matrix, a matrix of size $L \times L$, where L denotes the number of possible classes. This matrix is created by matching the predicted and target values. L was set to 2 (two) in the present case. Thus, having in mind to evaluate the performance of the proposed model the dataset was divided in exclusive subsets through the ten-folds cross validation [21]. In the implementation of the respective dividing procedures, ten accomplishments were performed for each one of them. Table 1 presents the coincidence matrix of the proposed model, where the values presented denote the average of 30 (thirty) experiments. A perusal to Table 1 shows that the model accuracy was 89.7% (i.e., 182 instances correctly classified in 203). Thus, the predictions made by the presented model are satisfactory, attaining accuracy close to 90%.

Table 1   The coincidence matrix for proposed model

Target	Predictive	
	True (1)	False (0)
True (1)	43	5
False (0)	16	139

Based on coincidence matrix it is possible to compute *sensitivity, speci-ficity, Positive Predictive Value (PPV)* and *Negative Predictive Value (NPV)* of the classifier. Briefly, *sensitivity* evaluates the proportion of true positives that are correctly identified as such, while *specificity* translates the proportion of true negatives that are correctly identified. *PPV* stands for the proportion of cases with positive results which are correctly classified while *NPV* is the proportion of cases with negative results which are successfully labeled [25, 26]. The values obtained for *sensitivity, specificity, PPV* and *NPV* were 89.6%, 89.7%, 72.9% and 96.5%, respectively. In addition, the *Receiver Operating Characteristic (ROC)* curve was considered. An *ROC* curve displays the trade-off between sensitivity and specificity. The *Area Under the Curve (AUC)* quantifies the overall ability of the test to discriminate between the output classes [25, 26]. Figure 11 depicted the *ROC* curve for the proposed model. The area under *ROC* curve close to 0.9 denoting that the model exhibits a good performance to signalize evidences of dyscalculia.

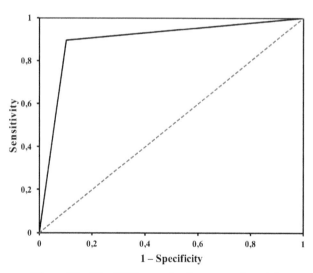

Figure 11   The *ROC* curve for the proposed model.

6 Conclusions

This work presents an intelligent decision support system to estimate children evidences of dyscalculia based on the use of *android app disMAT*. It is centred on a formal framework based on *LP* for *Knowledge Representation and Reasoning*, complemented with a *CB* approach to computing that caters for the handling of incomplete, unknown, or even self-contradictory information. The proposed model is able to provide adequate responses once the overall accuracy is close to 90% and the area under *ROC* curve is near 0.9. The methodology followed in this work may set the basis of an overall approach to such systems, susceptive of application in different arenas. Furthermore, under this line of thinking the cases' retrieval and optimization phases were heightened when compared with existing tactics or methods. Additionally, under this scenery the users may define the weights of the cases' attributes on the fly, letting them to choose the most appropriate strategy to address the problem (i.e., it gives the user the possibility to narrow the search space for similar cases at runtime).

References

[1] Kosc, L. (1974). Developmental dyscalculia. *J. Learn. Disabil.* 7, 164–177. doi: 10.1177/002221947400700309 PMID:NOPMID

[2] Berch, D., and Mazzocco, M. (2007). *Why Is Math So Hard for Some Children? The Nature and Origins of Mathematical Learning Difficulties and Disabilities.* Baltimore, MD: Paul H. Brookes Publishing Co.

[3] Ardila, A., and Rosselli, M. (2002). Acalculia and dyscalculia. *Neuropsychol. Rev.* 12, 179–231. doi: 10.1023/A:1020329929493 PMID:12539968

[4] Price, G. R., and Ansari, D. (2013). Numeracy advancing education in quantitative literacy dyscalculia: characteristics, causes, and treatments. *Numeracy* 6, doi: org/10.5038/1936-4660.6.1.2

[5] Geary, D. C. (1993). Mathematical disabilities: cognitive, neuropsychological, and genetic components. *Psychol. Bull.* 114, 345–362. doi: 10.1037/0033-2909.114.2.345 PMID:8416036

[6] Romagnoli, G. (2008). *Dyscalculia: A Challenge in Mathematics* (in Portuguese). São Paulo: CRDA.

[7] Barkley, R. (1982). "Guidelines for Defining Hyperactivity in Children," in *Advances in Clinical Child Psychology*, Vol. 5, eds B. Lahey and A. Kazdin (New York: Springer), 137–180.

[8] Michaelson, M. T. (2007). An Overview of Dyscalculia: Methods for Ascertaining and Accommodating Dyscalculic Children in the Classroom. *Aust. Math. Teach.* 63, 17–22. doi: NODOI PMID:NOPMID

[9] Rubinsten, O., and Henik, A. (2008). Developmental dyscalculia: heterogeneity might not mean different mechanisms. *Trends Cogn. Sci.* 13, 92–99. doi: 10.5363/tits.13.6_92 PMID:19138550

[10] Ferraz, F., and Neves, J. (2015). "A brief look into dyscalculia and supportive tools," in *Proceedings of the 5th IEEE International Conference on E-Health and Bioengineering (EHB 2015)* (Rome: IEEE), 1–4.

[11] Aamodt, A., and Plaza, E. (1994). Case-based reasoning: foundational issues, methodological variations, and system approaches. *AI Commun.* 7, 39–59. doi: NODOI PMID:NOPMID

[12] Richter, M. M., and Weber, R. O. (2013). *Case-Based Reasoning: A Textbook.* Berlin: Springer.

[13] Neves, J. (1984). "A logic interpreter to handle time and negation in logic databases," in *Proceedings of the 1984 annual conference of the ACM on the 5th Generation Challenge*, eds R. Muller and J. Pottmyer (New York: Association for Computing Machinery), 50–54.

[14] Kakas, A., Kowalski, R., and Toni, F. (1998). "The role of abduction in logic programming," in *Handbook of Logic in Artificial Intelligence and Logic Programming*, Vol. 5, eds D. Gabbay, C. Hogger, and I. Robinson (Oxford: Oxford University Press), 235–324.

[15] Pereira, L., and Anh, H. (2009). "Evolution prospection," in *New Advances in Intelligent Decision Technologies – Results of the First KES International Symposium IDT 2009*, Studies in Computational Intelligence, Vol. 199, ed. K. Nakamatsu (Berlin: Springer), 51–64.

[16] Neves, J., Machado, J., Analide, C., Abelha, A., and Brito, L. (2007). "The halt condition in genetic programming," in *Progress in Artificial Intelligence*, LNAI, Vol. 4874, eds J. Neves, M. F. Santos, and J. Machado (Berlin: Springer), 160–169.

[17] Lucas, P. (2003). "Quality checking of medical guidelines through logical abduction," in *Proceedings of AI-2003 (Research and Developments in Intelligent Systems XX)*, eds F. Coenen, A. Preece, and A. Mackintosh (London: Springer), 309–321.

[18] Machado, J., Abelha, A., Novais, P., Neves, J., and Neves, J. (2008). "Quality of service in healthcare units," in *Proceedings of the ESM 2008*, eds C. Bertelle and A. Ayesh (Ghent: Eurosis – ETI Publication), 291–298.

[19] Fernandes, F., Vicente, H., Abelha, A., Machado, J., Novais, P., and Neves, J. (2015). "Artificial Neural Networks in Diabetes Control," in *Proceedings of the 2015 Science and Information Conference (SAI 2015)*, (Rome: IEEE), 362–370.

[20] Vicente, H., Couto, C., Machado, J., Abelha, A., and Neves, J. (2012). Prediction of Water Quality Parameters in a Reservoir using Artificial Neural Networks. *Int. J. Design Nat. Ecodyn.* 7, 310–319. doi: 10.1016/j.scitotenv.2014.09.005 PMID:25241206

[21] Haykin, S. (2009). *Neural Networks and Learning Machines*. New Jersey: Pearson Education.

[22] Mendes, R., Kennedy, J., and Neves, J. (2003). "Watch thy neighbor or how the swarm can learn from its environment," in *Proceedings of the 2003 IEEE Swarm Intelligence Symposium (SIS'03)*, (Rome: IEEE), 88–94.

[23] Carneiro, D., Novais, P., Andrade, F., Zeleznikow, J., and Neves, J. (2013). Using Case-Based Reasoning and Principled Negotiation to provide decision support for dispute resolution. *Knowl. Inform. Syst.* 36, 789–826. doi: 10.1007/s10115-012-0563-0 PMID:NOPMID

[24] Figueiredo, M., Esteves, L., Neves, J., and Vicente, H. (2016). A data mining approach to study the impact of the methodology followed in chemistry lab classes on the weight attributed by the students to the lab work on learning and motivation. *Chem. Educ. Res. Pract.* 17, 156–171. doi: 10.1039/C5RP00144G PMID:NOPMID

[25] Florkowski, C. (2008). Sensitivity, Specificity, Receiver-Operating Characteristic (ROC) Curves and Likelihood Ratios: Communicating the Performance of Diagnostic Tests. *Clin. Biochem. Rev.* 29(Suppl. 1), S83–S87. doi: NODOI PMID:18852864

[26] Hajian-Tilaki, K. (2013). Receiver Operating Characteristic (ROC) Curve Analysis for Medical Diagnostic Test Evaluation. *Caspian J. Intern. Med.* 4, 627–635. doi: NODOI PMID:24009950

Biographies

F. Ferraz was born in Braga, Portugal and went to the University of Minho, where she studied Biomedical Engineering, and obtained her master degree in 2015. In 2016 she enrolled in the Doctoral Program in Biomedical Engineering, in the branch of medical informatics. Her master thesis and is theme of study in the Ph.D. are related to dyscalculia and its diagnosis and therapeutics. She is also a researcher in the Algoritmi Center of University of Minho, where she is developing software to Bosch in the line of industry 4.0. Her current interest regard Dyscalculia and Learning Disabilities, Neurology, Pediatrics, Psychiatrics, Medical Image, Internet of Things, Artificial Intelligence, Intelligent Systems, Data Mining, Knowledge Representation and Reasoning Systems, Computer Engineering, Software Engineering, Computer Science, Information Systems, and Information Technology.

H. Vicente was born in S. Martinho do Porto, Portugal and went to the University of Lisbon, where he studied Chemistry and obtained his degrees in 1988. He joined the University of Évora in 1989 and received his Ph.D. in Chemistry in 2005. He is now Auxiliary Professor at the Department of Chemistry at the University of Évora. He is a researcher at the Évora Chemistry Center and his current interests include Water Quality Control,

Lakes and Reservoirs Management, Data Mining, Knowledge Discovery from Databases, Knowledge Representation and Reasoning Systems, Evolutionary Intelligence and Intelligent Information Systems.

A. D. Costa, Ph.D., is an Assistant Professor at Department of Informatics, University of Minho, Portugal, where he develops teaching and research activities in the fields of Computer Networks and Computer Communications since 1992. As a researcher, he currently integrates the Computer Communications and Networks (CCN) research group, at Centro Algoritmi, School of Engineering, University of Minho. He graduated in Systems and Informatics Engineering in 1992, obtained a M.Sc. Degree in Informatics in 1998 and a Ph.D. Degree in Computer Science in 2006 at the same university. He participated in several research projects, supervised M.Sc. and Ph.D. students, and co-authored more than thirty peer reviewed papers in Routing Protocols, Network Services, Quality of Service, P2P, IoT and Network Management. His current research interests are in Mobile AdHoc Networks, Disruptive Delay Tolerant Networks, Named Data Networks, Indoor Positioning, Internet of Things and Future Internet.

J. Neves is Full Professor of Computer Science at Minho University, Portugal, since 1983. Jose Neves is the Deputy Director of the Division for Artificial Intelligence (AI). He received his Ph.D. in Computer Science from Heriot Watt University, Scotland, in 1983. His current research interests relate to the areas of Knowledge Representation and Reasoning, Evolutionary Intelligence, Machine Learning, Soft Computing, aiming to construct dynamic virtual worlds of complex symbolic entities that compete against one another in which fitness is judged by one criterion alone, intelligence, here measured in terms of a process of quantification of the quality of their knowledge, leading to the development of new procedures and other metaheuristics and their application in complex tasks of optimization and model inference in distinct areas, namely in the healthcare arena (e.g., machine learning in an intensive care unit environment).

Edge Resource Utilization Using OS-Level Virtualization on Smart Devices*

Thomas Renner**, Marius Meldau
and Andreas Kliem

Technische Universität Berlin (TU Berlin),
Complex and Distributed IT Systems (CIT),
10587 Berlin, Germany
***Corresponding Author: thomas.renner@tu-berlin.de*

Received 19 August 2016; Accepted 17 October 2016;
Publication 2 November 2016

Abstract

Following the Internet of Things (IoT) paradigm, more and more connected and embedded devices like sensors, smart phones, or entertainment devices are becoming available surrounding us. This increasing number of IoT devices leads to an increasing amount of resources made available to the users. For instance, if devices like smart phones or smart TVs are considered, this includes computing and storage resources. In this paper we propose a container-based resource allocation scheme to increase the utilization of these IoT resources and made them shareable between different users due to cloud-centric IoT deployments. The approach allows various applications and users to dynamically allocate resources offered by edge devices and process IoT data close to the source. The approach is evaluated regarding its feasibility in terms of performance on a exemplary resource constrained IoT device.

Keywords: Internet of Things, Edge Computing, OS-level Virtualization, Resource Management, Resource Sharing.

---

*An earlier version of this paper appeared in the Proceedings of International Conference on Software Networking (ICSN), IEEE, 2016.

Journal of Software Networking, 79–100.
doi: 10.13052/jsn2445-9739.2016.005

1 Introduction

The Internet of Things (IoT) paradigm [1] is based on the assumption that uniquely identifiable things can be integrated into information networks. Therefore, a virtual representation of the physical things needs to be created. This task can be accomplished by circuits embedded or attached to the things or by IoT Integration Platforms, such as a middleware running on a gateway device. In general, a virtual representation can be described as an interface that exposes the capabilities of the thing using a set of operations (e.g. getters or setters allowing to access certain properties). Once virtual representations are available, they can be integrated into information networks, such as the Internet, by means of communication. Thus, enabling developers to integrate the things into business, information and social processes (i.e. workflows and applications). As a result, the IoT paradigm can be envisioned as an enabler for a variety of applications found in consumer and industry related application domains. Popular examples are Smart Homes, Smart Cities, (Factory) Automation, E-Health, or Logistics [19, 32].

In order to allow application developers to properly take advantage of all the interconnected things, providing the IoT Integration Platform is the key challenge for most of the IoT deployments. Due to its various application domains, the IoT subject to a huge heterogeneity in terms of things, their virtual representations and the components required to create these representations (e.g. device drivers, data formats, configurations). These often follow different or even no standards [26]. As a result, IoT Integration Platforms need to hide the process of establishing physical integration and spontaneous interoperability from the application developers. This challenge is intensified by the rapid growth of things deployed to our environment, which is motivated by the proliferation of low cost, small size and more powerful devices [28] and by the increasing attention IoT based solutions gain.

Many solutions for IoT Integration Platforms currently available and discussed in literature follow a Cloud-centric approach. A Cloud-centric approach, sometimes referred to as Sensor Virtualization, basically consists of an edge node acting as a gateway (e.g. a router or a smart phone) for the things and a set of Cloud nodes providing the resources to perform data analytics and host applications. Thereby, the gateway node physically integrates the things and dispatches the resulting data streams to the Cloud nodes (i.e. takes care of Machine-to-Machine (M2M) [20] related aspects of the solution). Application logic and major parts required to achieve interoperability (e.g. data transformation, unification of interfaces) are entirely shifted to the Cloud. As a result, Cloud-centric solutions separate data acquisition and processing from each other. This causes the underlying network infrastructure to meet high requirements regarding non-functional properties. Low latency, high bandwidth, and efficient security and privacy measures are required to meet the Quality of Service expectations of IoT applications, especially in terms of applications related to the Industrial IoT domain [18]. Furthermore, the network infrastructure needs to provide sufficient resources in order to avoid congestion in the presence of the continuously growing amount of connected things.

Various initiatives and approaches like Fog Computing [5] or Mobile Edge Computing [25] have recognized the demand for more flexible IoT Integration Platforms and propose utilizing resources at the edge of the network. The objective is to reduce the latency and the amount of data that need to be transmitted between data sources and sinks (e.g. applications or consumers). The gap between data acquisition and processing resulting from Cloud-centric approaches can be bridged by moving application logic (e.g. (pre-) processing) towards the gateway nodes of the IoT Integration Platform or the things themselves. Several applications are proposed to benefit from the reduced latency (e.g. traffic control systems). A further motivation of this approach is given considering the growing amount of so called Smart Devices. Examples for Smart Devices include smart phones, smart TVs, game consoles, or routers, which are already used as gateways in IoT integration platforms. Basically, a Smart Device can be distinguished from a usual sensor or thing by its capability to execute custom software components. In addition to its role as a data source in terms of the IoT, a Smart Devices offers computing and storage capabilities. Hence, internal resources like CPU or storage can be made available to external components such as Apps or software components for data (pre-) processing. By facilitating these resources, the IoT Integration Platform can be enriched with data processing capabilities available at the edge of the network, which means close to the data sources (and possibly consumers as well). However, most of these Smart Devices are currently used for stand-alone purposes and need to be enriched with capabilities to share their processing power for instance. A suitable example is given by a smart TV. Current smart TVs are based on a fully featured computing platform to offer services like video streaming, internet access or other entertainment services. However, often these TVs are only used for a few hours per week and remain idle the meantime.

With this motivation, we propose to apply the popular Cloud Computing paradigm [21] to the IoT domain. We consider things as resources that may offer their computing or storage capabilities. These can be allocated on demand independent from the owning user. This introduces the notion of sharing IoT devices and the capabilities they offer. As a result, the utilization of compute and storage capabilities can be increased and the communication demand and latency can be reduced, because edge devices are used for data (pre-) processing and aggregation. In addition, a resource oriented view on the available IoT infrastructure enables new ways of mobility, scalability, and upcoming paradigms like community based interaction and sharing of IoT resources [12, 14].

This paper presents an approach towards a container-based resource management system for connected, embedded IoT devices. The approach is based on the Device Cloud [29], an IoT architecture that focuses the application of Cloud Computing paradigms to the IoT domain by enabling users to share and allocate physical IoT devices using a resource-pool abstraction. The resource management capabilities of the presented approach are based on Docker [22]. We propose a container-based virtualization approach levering the advantages of the underlying Operating-system-level (OS-level) virtualization. OS-level virtualization provides sufficient resource isolation capabilities and provides less overhead compared to a virtual-machine based approach [31]. Appropriate isolation is a mandatory requirement, since the proposed approach assumes that resources of a single Smart Device can be allocated and shared by multiple users. In particular, we introduce an architecture that takes advantage of Docker's automated deployment model and allows sharing resources within containers between different users and devices. The presented approach focuses the feasibility of OS-level virtual-ization for the isolation of user code and the dynamic allocation of resources offered by Smart Devices.

The remainder of the paper is structured as follows. Section 2 discusses approaches for the isolation of user code by means of virtualization on resource constrained devices. Section 3 explains our approach and system architecture. Section 4 presents first results, including our prototype and an evaluation of it. Section 5 discusses further related work Section 6 concludes this paper.

2 Background

Virtualization can be considered as one of the core enablers leading to the success of the Cloud Computing paradigm. It allows providers of computing infrastructure to provision currently idle resources and thus, can significantly benefit the utilization of resources. Its main advantage is the high level of isolation achieved between tasks belonging to different users. However, with respect to mobile devices, which usually suffer from resource constraints and regard battery life-time as one of their most expensive resources [4], the efficiency of the virtualization technique employed becomes of crucial importance.

The most comprehensive level of isolation is achieved by full (hardware) virtualization. A hypervisor provides a full set of virtual devices to guest operating systems. As a result, guest operating systems usually do not need to be aware of the virtualization and can be operated in the virtual environment without modification. However, because full virtualization requires to pass all hardware accesses through virtualized drivers, these solutions are constrained by reduced I/O performance and considerable virtualization overhead. Since resource constraints play an important role for embedded devices, paravirtualization has been considered as a virtualization approach for such environments. Paravirtualization replaces the virtual I/O-stack with a more efficient and lightweight API.

The most popular paravirtualization solution is the open source software Xen [2]. The concept is to run a privileged virtualization domain called *dom0* that has access to the underlying hardware and multiple limited domains called *domU* that access the hardware via *dom0*. While the *dom0* can rely on native device drivers to integrate the underlying hardware, the *domU* domains require Xen-specific paravirtualization drivers to access the hardware through the *domU*. Xen is lightweight enough to run on a smart device [8]. Based on the Xen Project, embeddedXEN [30] is a project specifically aimed at running the Xen hypervisor on smart devices. It uses a multi-kernel binary image which can be flashed for example on a smartphone. This image makes deployment relatively easy. However, the specific architecture employed by the embeddedXen project limits Xen's capabilities to only a single *domU* besides the obligatory *dom0*. In general, the installation and deployment of Xen on smart devices introduces some difficulties. The main challenge is to provide the required paravirtualization drivers. While sensors may be integrated using a trusted gateway *domU*, each *domU* would still need to be able to access the hardware resources offered by the host devices (e.g. storage). Since the amount of different hardware platforms available for smart devices is considerably greater than for regular server or PC platforms, providing specialized paravirtualization drivers for each possible platform may not be feasible.

Another virtualization solution available for smart devices is the VMware Mobile Virtualization Platform (MVP) [3], which is a type-2 paravirtualizing hypervisor hosted on Android and an ARM processor. The major objective of MVP is to enhance corporate IT-infrastructures with Bring Your Own Device (BYOD) capabilities. BYOD allows employees to use their private devices and integrate them securely into a corporate IT environment following certain policies and conditions (e.g. isolation of user and company data). The MVP does not place any restrictions on the host system, but creates virtual guest environments that can be secured by the IT department. However, MVP still suffers from the requirement to provide paravirtualization drivers.

2.1 OS-Level Virtualization

Operating System (OS) Containers or Operating-System-Level Virtualization is a completely different isolation concept. It does not provide virtual hardware environments like full and paravirtualization. Rather, it adds an additional layer of isolation within the Kernel of the host OS. This yields multiple virtualized userland environments, each of which has a view on the system as if it was running alone on the machine [16].

Recently, OS-Level virtualization is gaining attention as the concept behind the scenes of the Docker platform [7]. Although recently replaced by another API, Docker was initially developed on top of the Linux Containers (LXC) library. Thus, Docker allows isolating applications by means of Container-based OS-level virtualization. A Container provides all necessary dependencies an application requires. Furthermore, the Linux Kernel features control groups (cgroups) and namespaces allow controlling the resource consumption of a particular Container and isolating the Containers from each

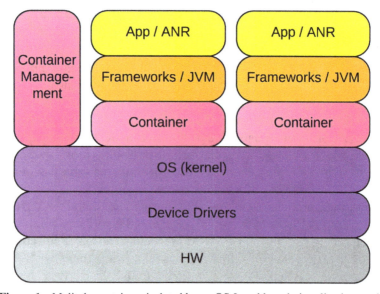

Figure 1   Mulitple containers isolated by an OS-Level based virtualization stack.

other. In comparison to full and paravirtualized approaches, OS-level virtualization (i.e. Container virtualization) features are provided by the Kernel without the need to operate a distinct hypervisor. Therefore, it is considered as more lightweight and performant. Although providing less resource isolation capabilities than full hardware or paravirtualization, we consider the level of isolation as sufficient for the IoT resource management scenario.

Docker defines a scripted build process for container images similar to the classical build process of software. It also uses an online registry where those images can be published. Other images can be build based on those images, and finally these images can be started as actual containers on a Docker host. By using differential-based / copy-on-write file systems, a fast distribution of updates can be achieved. Thus, additionally to its virtualization capabilities, Docker can be considered as a version control system for binary images. Therefore, Docker features an efficient distribution scheme appropriate to implement and deploy *Task Execution Containers* among the smart device platforms that may be used in the IoT resource management scenario.

The efficiency of Docker's facilities for storing and distributing images of Containers is based on a flexible and exchangeable storage backend. By default Docker uses the overlay file system AUFS. AUFS combines several directories into a virtual file system with multiple layers, called branches. The lower layers are read only, but files from these layers can logically be changed by (re)writing a file on the topmost branch [23]. Also, there are backends available for devicemapper which uses virtual block devices [17], and the copy-on-write file system btrfs [27]. This means that even large amounts of *Task Execution Containers* can be maintained with little overhead compared to a single one.

Besides deploying images, Docker provides a registry, which basically acts as a pool of repositories containing Docker images. It allows developers to publish new and updated images and therefore offers an already exiting approach for managing the set of *Task Execution Containers* required to set up the resource management scenario on different smart device platforms.

3 Resource Sharing Architecture

This section describes the design of the Container-based approach we propose to manage and share resources of IoT devices between different users and applications. As outlined in Section 1, scalability is an important requirement to meet the growing demands of the users and handle the growing amount of deployed IoT devices and their related resources. Therefore, the IoT Resource

Management Framework presented in this section is designed relying on a federated approach, introducing several collaborating domains. After describing the overall concept of the domain network, the composition of a single domain network in terms of its nodes is discussed. Finally, life cycle operations required to deploy, initialize and release nodes and *Task Execution Containers* are introduced.

3.1 Device Cloud Domain Network

The Device Cloud follows a federated approach to tackle the scalability challenge. It consists of a set of Domains, whereas each Domain may entirely operate on its own without being required to collaborate with any other Domain. This is based on the assumption that resource sharing interactions usually stay within a certain group of users and therefore follow the locality principle. Therefore, each Domain provides a set of management capabilities, that enable nodes to interact with each other (e.g. directory services or the *Task Execution Containers* repository). These management capabilities are hosted by an independent Domain Operator not participating in the resource sharing processes (similar to the DNS system). However, resource sharing is not restricted to the set of devices contained in one Domain. As described by the general concept of sharing devices for the purpose of increasing resource utilization [12], interactions crossing Domain boundaries are possible. Therefore, the Domain Operators need to negotiate access to so called remote resources on behalf of the users be part of their Domain. As shown in Figure 2, a Domain basically consists of different types of nodes types, whereas no assumptions on the topology are made:

- **Aggregation Nodes (ANs):** ANs represent the smart devices that build the foundation of the proposed resource sharing approach. We use the term Aggregation Node because these nodes are able to conduct generic computational tasks like, for instance, aggregation or pre-processing of data. In general, ANs are devices that provide sufficient compute- and storage resources to run multiple containers without compromising the device main functionality, such as the ability to receive phone calls. Typically, a Domain consists of many interconnected ANs.
- **Backend Nodes (BNs):** BNs represent nodes that are used to host the management capabilities of a Domain. Thus, BNs are usually operated by the Domain Operated and are likely to be deployed on dedicated servers in a data center. BNs provide directory, monitoring and management

services to ANs. For instance, BNs are responsible for scheduling containers and distribute Docker images among ANs.

- **Device Nodes (DNs):** DNs are sensors or things that do not fulfill the AN characteristics and thus only act as data sources (e.g. sensors). DNs are usually integrated into the Domain information network by ANs. However, since DNs do not apply for the Docker based resource management, these nodes are not shown in Figure 2 and their integration is not further discussed.

3.2 Aggregation Node Architecture

According to the node classification given in the previous section, ANs constitute the target for the OS-level based virtualization that is required to allow sharing resources of an AN among multiple users. This section will describe the architecture of a single AN. We assume that the platform of the

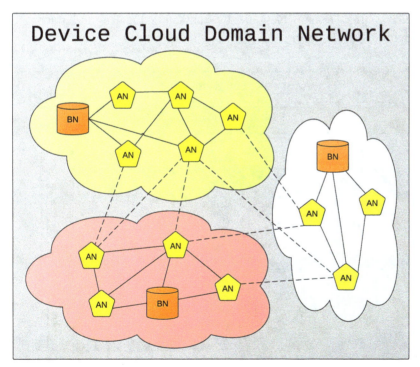

Figure 2 Federated device cloud domains consisting of interconnected Aggregation Nodes (AN) and Backend Nodes (BN).

AN (e.g. Android) is Linux based and supports all Kernel features required to run Docker.

Figure 3 illustrates the internal architecture of an AN. Besides the Linux Kernel and the corresponding Operating System hosted on the AN, following components and containers are defined for the purpose of making the (computational) resources of the node available:

- **Bootstrap Code**: The bootstrap code is necessary to set up the Docker based resource sharing environment on an AN. It retrieves the most recent container images from its corresponding BN and allocates containers inside the Docker Daemon. The Docker Deamon can be also referred to as the Aggregation Node Runtime, which manages our *Task Execution Containers*.

- **Management ANR**: The Management ANR is a singleton container that provides the management facilities required within the Aggregation Node Runtime. This includes life-cycle operations required to start, stop, or update containers. Furthermore, the Management ANR registers the AN with the Domain Network by contacting the BN and starting a neighbor AN discovery. As a result, the resources of the AN can be

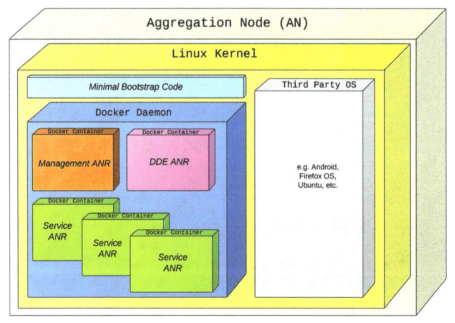

Figure 3   Overview of the aggregation node architecture.

considered for all future resource allocation attempts. In order to keep track of the available resources, the Management ANR monitors the resource consumption of its running containers and reports the results to its corresponding BN for the purpose of container scheduling decisions.

- **Device Driver Engine (DDE) ANR**: Similar to the Management ANR, the DDE ANR is a singleton Container providing fundamental system services to the Aggregation Node Runtime. It is responsible for the physical integration and management of IoT devices and sensors and makes them available to other Containers. This includes local sensors embedded into the AN (e.g. a GPS sensor integrated into a smart phone) as well as devices connected through the communication interfaces of the AN (e.g. via Bluetooth). In general, the DDE is a framework for on-demand composition of device control logic and allows loading required integration knowledge (e.g. device drivers) from BNs at runtime. The DDE has been described in [13]. Given the DDE, application containers do not need to take care of the physical integration process of sensors and other devices. The DDE exposes them using a Device Category based interface format as described in [24]. Therefore, the BN is able to match the set of Device Categories available on an AN to the requirements of application containers and consider the results during its scheduling decisions.

- **Service ANR**: The Service ANR represents one instance of a *Task Execution Container.* Service ANRs are Docker containers that execute external application logic (i.e. user code) on the AN. Hence, Service ANRs make the resources of an AN available. By instantiating new Service ANRs, available resources can be allocated on-demand. Based on the type of the user code, different images for Service ANRs can be deployed. as outlined, each Servcie ANR can access external data sources, such as sensors, through the IPC interfaces offered by the DDE ANR.

Since the resources offered by a single smart device are often limited and may not be sufficient to fulfill the overall requirements of an application, several Service ANRs hosted on different ANs may be allocated for one application. Since these Service ANRs are likely to communicate with each other, a virtual network for each application (i.e. service) is established. As a result, two kinds of virtual networks exist within one Domain Network:

- **Management Network**: Each Domain includes one management net-work. The management network connects the Management ANRs and is used to exchange state information and control commands.

- **Service Network**: Each Domain includes a number of Service Networks, whereas the quantity depends on the number of applications active within the Domain. Each application is assigned its own virtual network. Using the Service Network, Service ANRs belonging to the same application can exchange application data.

3.3 Resource Sharing Life-Cycle

In order to allocate and release resource properly, certain life-cycle operations are defined for allocation and utilization of AN resources by applications. The operations basically describe the process of ANR container deployment, instantiation and release as well as the respective state transitions. As shown in Figure 4, the node and container initialization process consists of following steps:

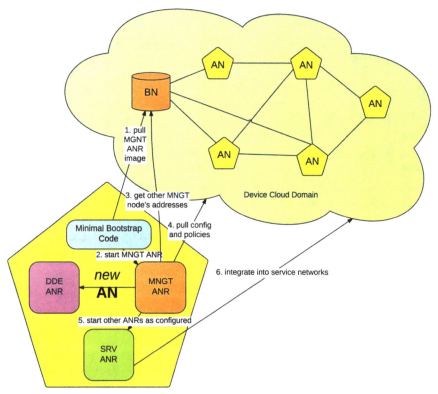

Figure 4   Node and container initialization. A new AN joins the device cloud network and an application is started.

1. **Pull Management ANR**: Upon initialization of an AN, the bootstrap code loads the most recent image of the Management ANR from the BN of its corresponding Domain Network.
2. **Start Management ANR**: Afterwards, the bootstrap code starts the Management ANR container, which now takes over management of the Aggregation Node Runtime.
3. **ANR Discovery**: On startup the Management ANR registers itself with the BN and starts a neighbor discovery to find all other ANs. Therefore, it uses the BN as an anchor node and retrieves a list of existing Management ANRs.
4. **Pull Configuration**: After having registered with the Domain Network, the Management ANR queries the network for a current set of policies and configurations. These can be retrieved from the BN or from adjacent ANs, which reduces possible delay due heavy load of the BN.
5. **Start DDE ANR**: Next, the Management ANR starts the DDE ANR in order to discover and integrate available DNs (e.g. sensors).
6. **Service Registration**: Application developers willing to facilitate the resources of a Domain, must upload their Service ANR containers to the BN. The BN takes care of scheduling the application by identifying possible ANs through the Management Network.
7. **Service Deployment**: Management ANRs being elected to host Service ANRs pull the recent container image from the BN and start up the container. The resulting Service ANR will announce itself to its corresponding Service Network.
8. **Shutdown**: Upon shutdown of the AN, the Management ANR asks all Service ANRs and the DDE ANR to stop and release allocated resources as well as terminate possible connections to other nodes. It then stops all running containers. Finally, the Management ANR requests the bootstrap code to terminate itself.

4 Evaluation

This section presents our experimental setup, the test workload, and benchmark results for a container-based Resource Management Platform. Since this paper focuses on the feasibility of the OS-level virtualization based resource allocation and isolation, it particularly targets performance measures of life cycle operations required to deploy and release containers. In addition, the computational overhead of executing applications in containers is measured by using a test application. All experiments were done using on a Raspberry

Pi 2 Model B with a 900MHz quad-core ARM Cortex-A7 CPU and 1GB RAM. The Raspberry Pi runs on Hypriot Docker for Raspberry Pi version 0.2[1], a Debian based operating system including a Docker package for ARM architectures.

4.1 Performance of Life-Cycle Operations

An important performance aspect of the resource management platform arises from life-cycle actions, required to manage the deployment of single ANR containers (e.g. init, start, stop). Naturally, these operations take some time and can have serious impact on the overall performance of the platform. This is because a Domain Network is supposed to exhibit a dynamic nature, which means frequent changes of service allocations and container deployments are subject to single ANs. Therefore, it is important to observe how fast the life-cycle transitions can be conducted on actual hardware. It is worth mentioning that we assume that the device on which the containers are running has already registered with the domain. Thus, the evaluation focuses on the duration of the following life-cycle operations:

- **Initialize:** Initializing an ANR requires to create a data-container and a corresponding runtime container.
- **Start:** After initialization, the runtime container is automatically started by invoking the entry-point defined in the Dockerfile corresponding to the runtime-container (e.g. a command to start user code).
- **Stop:** Stops the runtime container.
- **Update:** Stops a runtime container, loads the most recent image from a private Docker Hub and starts the container again.
- **Reset:** Stops and removes the runtime and data container.

Table 1 shows the time in seconds that a life-cycle action took per container. Three series of tests have been conducted, whereas one 1, 5 and 15 containers have been started in a sequential order. Every experiment was executed

Table 1   Execution times of different life cycle action in seconds for successive starting of 1, 5 and 15 containers

# Containers	Init	Start	Stop	Update	Reset
1	1.76	0.87	0.83	2.00	1.33
5	2.03	1.01	0.93	1.94	1.37
15	2.18	1.02	0.82	2.08	1.37

---

[1]http://blog.hypriot.com/, accessed 2016-04-01

seven times. We report the median runtime. The results shows that all life-cycle operations complete within an acceptable amount of time. With an increasing load, the average life cycle execution time per container increases by about 20%.

4.2 Computation Overhead

This subsection evaluates the computational overhead in terms of CPU utilization and the resulting impact on tasks running inside the containers. Therefore, the CPU performance was measured using the Scimark 2.0 benchmark[2]. This benchmark was developed for scientific and numerical computing. It measures several computational kernels and reports a composite score in approximate Mflop. To measure the performance of a task running in Docker, the Scimark jar file is treated as a piece of user code to be deployed inside a Service ANR. Therefore, the benchmark was executed upon startup on the respective container. For every run, the number of simultaneously running Scimark containers has been increased to simulate high system load. Table 2 shows the results of these experiments. As mentioned, the Raspberry Pi 2 used in our evaluation has four CPU cores. Scimark 2.0 is a single threaded application. Therefore, a relative score variable, expressing the score per instance by $\frac{no. \ of \ instances}{no. \ of \ cores}$, was introduced in Table 2. Thus, a value of 1.0 stands for an approximately equal performance result compared to a single benchmark run without using a Docker based container. The results show no relevant difference between native and dockerized benchmark processes, nor do they show a relevant drop in performance due to the scheduling overhead introduced due to many benchmark instances running simultaneously. As a result, the computational overhead can been assessed as low. Even 20 instances with heavy CPU usage slow down the overall performance only by 4%.

Table 2 Results of Scimark 2.0 in Mflop. Relative score is normalized on the number of cores, and relative to the native performance

# Containers	Composite Score	Relative Score
Bare Metal (Baseline)	14.19	–
1 Container	14.22	1.00
2 Containers	14.21	1.00
4 Containers	14.14	1.00
8 Containers	6.98	0.98
20 Containers	2.73	0.96

[2]http://math.nist.gov/scimark2/, accessed 2016-04-01

5 Related Work

This section presents related approaches that target the utilization of computational resources offered by edge devices.

Cloud computing paradigms like pooling and allocating a subset of the resources to certain tasks on-demand have recently been applied to sensor networks. Motivated by the increasing set of capabilities of single nodes, Virtual Sensor Networks (VSNs) assume that the overall set of nodes in a Wireless Sensor Network (WSN) can be virtually grouped into subsets and linked to specific applications or tasks [9, 10]. Unlike traditional WSNs, where usually all nodes perform similar tasks, a subset of nodes can be allocated for a given time period to perform specific tasks based on the observations of the environment (e.g. a critical event was recognized).

Another approach exploiting virtualization techniques for sensor nodes is called Cyber-Physical Cloud Computing [6, 11]. By applying lightweight virtualization capabilities to sensor nodes, sensors basically act as servers that move in space and execute virtual sensors. Virtual sensors can migrate between physical ones, which is referred to as cyber-mobility (i.e. moving between sensor hosts). Additionally, virtual devices can move with their current sensor host (if the sensor node is mobile), which is referred to as physical mobility.

Krainer et al. [15] introduce an idea for a platform as a service (PaaS) solution that uses physical vehicle swarms in a virtualized manner to provide information acquisition services. The authors propose a virtual vehicle monitor (VVM), similar to a virtual machine monitors for server virtualization. This VVM creates a set of virtual vehicles (VV) that can be configured via a web interface. At the same time the VVM controls the physical vehicles and manages the virtual vehicle software agents. It is also responsible for the migration of software agents between physical vehicle nodes. Clients can get access to several virtual vehicles via the web interface, and program a complex information acquisition task. A use case is given by surveying the humidity and temperature in a forest to assess the risk for fires. This task is then mapped by the VVM to the physical vehicles considering the requests from other VVs.

6 Conclusion

This paper proposes a container-based approach for an Internet of Things Resource Management Platform. The approach is evaluated regarding its feasibility to run on resource-constrained Smart Devices. Using OS-level based virtualization, applications can dynamically allocate containers on available smart devices such as smart phones or smart TVs connected to their network and utilize the offered resources more efficiently. Furthermore, the approach allows sharing the resources between multiple users and applications. The presented prototype and evaluation based on Docker shows that OS-level virtualization is a reasonable choice in terms of computational overhead and deployment time.

Future work will primarily target the improvement of our approach regarding memory and storage resources. In addition, container monitoring and scheduling strategies that take the characteristics of our presented Device Cloud Network, such as high mobility, into account, need to be investigated.

References

[1] Ashton, K. That internet of things thing. (2009). *RFiD J.* 22, 97–114.

[2] Barham, P., Dragovic, B., Fraser, K., Hand, S., Harris, T., Ho, A., Neugebauer, R., Pratt, I., and Warfield, A. (2003). Xen and the art of virtualization. *ACM SIGOPS Operati. Syst. Rev.* 37, 164–177.

[3] Barr, K., Bungale, P., Deasy, S., Gyuris, V., Hung, P., Newell, C., Tuch, H., and Zoppis, B. The vmware mobile virtualization platform: is that a hypervisor in your pocket? *SIGOPS Oper. Syst. Rev.* 44, 124–135.

[4] Bell Media. *Battery Life Trumps All Other Smartphone Features for Consumers: Survey.* Available at: http://www.ctvnews.ca/sci-tech/battery-life-trumps-all-other-smartphone-features-for-consumers-survey-1:1821 084 (Accessed September 20, 2016).

[5] Bonomi, F., Milito, R., Zhu, J., and Addepalli, S. (2012). "Fog computing and its role in the internet of things," in *Proceedings of the First Edition of the MCC Workshop on Mobile Cloud Computing, MCC '12*, 13–16 (New York, NY: ACM).

[6] Craciunas, S. S., Haas, A., Kirsch, C. M., Payer, H., Röck, H., Rottmann, A., Sokolova, A., Trummer, R., Love, J., and Sengupta, R. (2010). "Information-acquisition-as-a-service for cyber-physical cloud computing," in *Proceedings of the 2nd USENIX conference on Hot topics in cloud computing*, 14–14. (Berkeley, CA: USENIX Association).

[7] Docker Inc. (2015). "Understanding Docker," in *Docker Documentation* (online February 24, 2015).

[8] Hwang, J.-Y., Suh, S.-B., Heo, S.-K., Park, C.-J., Ryu, J.-M., Park, S.-Y., and Kim, C.-R. (2008). "Xen on ARM: system virtualization using Xen hypervisor for ARM-based secure mobile phones," *in 2008 5th*

IEEE Consumption Communication Network Confence, 257–261 (Rome: IEEE).

 [9] Islam, M. M., Hassan, M. M., Lee, G.-W., and Huh, E.-N. (2012). A survey on virtualization of wireless sensor networks. *Sensors*, 12, 2175–2207.

[10] Jayasumana, A. P., Han, Q., and Illangasekare, T. H. (2007). "Virtual sensor networks – a resource efficient approach for concurrent applications," in *Fourth International Conference on Information Technology (ITNG '07)*, 111–115.

[11] Kirsch, C., Pereira, E., Sengupta, R., Chen, H., Hansen, R., Huang, J., Landolt, F., Lip-pautz, M., Rottmann, A., Swick, R., Trummer, R., and Vizzini, D. (2012). "Cyber-physical cloud computing: the binding and migration problem," in *Design, Automation Test in Europe Conference Exhibition (DATE)*, 1425–1428.

[12] Kliem, A., and Kao O. (2015). "The internet of things resource management challenge," in *2015 IEEE International Conference on Data Science and Data Intensive Systems*, 483–490 (Rome: IEEE).

[13] Kliem, A., Koener, M., Weissenborn, S., and Byfield, M. (2015). "The device driver engine-cloud enabled ubiquitous device integration," in *2015 IEEE Tenth International Conference on Intelligent Sensors, Sensor Networks and Information Processing (ISSNIP)*, 1–7 (Rome: IEEE).

[14] Kliem, A., and Renner, T. (2015). "Towards on-demand resource provisioning for iot environments," in *Intelligent Information and Database Systems*, 484–493 (Berlin: Springer).

[15] Krainer, C., and Kirsch, C. M. (2014). "Cyber-physical cloud computing implemented as paas," in *Proceedings of the 4th ACM SIGBED International Workshop on Design, Modeling, and Evaluation of Cyber-Physical Systems*, 15–18. (New York, NY: ACM).

[16] Laadan, O., and Nieh, J. (2010). "Operating system virtualization: practice and experience," in *Proceedings of 3rd Annual Haifa Exp. System Conference* (New York, NY: ACM).

[17] Larsson, A. (2016). *devicemapper – a storage backend based on Device Mapper*.

[18] Laukkarinen, T., Suhonen, J., and Hannikainen, M. (2013). An embedded cloud design for internet-of-things. *Int. J. Distribut. Sensor Netw*. 9, 790130.

[19] Libelium Comunicaciones Distribuidas S.L. (2014). *Top 50 Internet of Things Applications – 50 Sensor Applications for a Smarter World*. Available at: http://www.libelium.com/top_50_iot_ sensor_appli cations_ranking/ (accessed April 01, 2016).

[20] M2M Alliance e.V. (2007). *Machine-to-Machine (M2M) - Whitepaper.* Available at: http://www. m2m-alliance.de/uploads/media/Whitepaper. pdf (accessed April 01, 2016).

[21] Mell P., and Grance, T. (2011). The nist definition of cloud computing. Available at: available at: http://dx.doi.org/10.6028/NIST.SP.800-145

[22] Docker, D. M. (2014). *Lightweight linux containers for consistent development and deployment*, Linux J.

[23] Okajima, J. R. (2015). *Aufs4 – advanced multi layered unification filesystem version 4.x.*

[24] OSGi Alliance. (2016). *OSGi Compendium Release 5 - Device Access Service Specification.* Technical report (accessed September 22, 2016).

[25] Patel, M., Naughton, B., Chan, C., Sprecher, N., Abeta, S., Neal, A., et al. (2014). *"Mobile-edge computing introductory technical white paper,"* in White Paper, Mobile-edge Computing (MEC) Industry Initiative.

[26] Perera, C., Zaslavsky, A., Christen, P., and Georgakopoulos, D. (2014). Context aware computing for the internet of things: a survey. *IEEE Commun. Surv. Tutorials* 16, 414–454.

[27] Petazzoni, J. (2013). "BTRFS support for Docker," in *Docker Blog.*

[28] Raj Rajkumar, R., Lee, I., Sha, L., and Stankovic, J. (2010). "Cyber-physical systems: the next computing revolution," in *Proceedings of the 47th Design Automation Conference*, 731–736 (New York, NY: ACM).

[29] Renner, T., Kliem, A., and Kao, O. (2014). "The device cloud-applying cloud computing concepts to the internet of things," in *Proceedings of the 2014 IEEE 11th International Conference on Ubiquitous Intelligence and Computing*, 396–401 (Washington, DC: IEEE Computer Society).

[30] Rossier, D. (2012). EmbeddedXEN: A Revisited Architecture of the XEN hypervisor to support ARM-based embedded virtualization.

[31] Soltesz, S., Pötzl, H., Fiuczynski, M. E., Bavier, A., and Peterson, L. (2007). "Container-based operating system virtualization: a scalable, high-performance alternative to hypervisors," in *ACM SIGOPS Operating Systems Review*, Vol. 41, 275–287 (New York, NY: ACM).

[32] Sundmaeker, H., Guillemin, P., Friess, P., and Woelfflé S (eds). (2010). *"Vision and challenges for realising the Internet of Things"*, in The Cluster of European Research Projects on the Internet of Things, Vol. 20 (Luxembourg: EUR-OP).

Biographies

T. Renner is postdoctoral student at the Complex and Distributed IT Systems Group of Technische Universität Berlin (TUB) in Berlin, Germany. His research interests include scalable data analytics, middleware-platforms for the Internet of Things, and building distributed systems.

M. Meldau achieved his master's degree in computer science at Technische Universität Berlin (TUB), Germany, in 2015. In his master's thesis, he designed and implemented a resource sharing concept for mobile device clouds.

A. Kliem is a postdoctoral researcher at the Complex and Distributed IT Systems Group of Technische Universitat Berlin (TUB) in Berlin, Germany. His current research interests center around middleware-platforms for the Internet of Things.

Modeling Voice over LTE

Ahmed Amou El Arby and Ousmane Thiare

University Gaston Berger, Saint Louis, Senegal
E-mail: elarby.amou@gmail.com; othiare@ugb.edu.sn

Received 15 September 2016; Accepted 11 October 2016;
Publication 22 November 2016

Abstract

The 4G technology, known as Long Term Evolution (LTE), has dominated the telecommunication market the last few years due to its high data rate (up to 100 Mbps) that will perfectly satisfy the users multimedia applications needs in term of bandwidth and that rely on IP domain network. However, LTE lack in offering services for applications that rely on Circuit Switched (CS) domain particularly the voice CS service. This paper studies the latest up to date solutions for the voice over LTE problem such as Single Radio Voice Call Continuity, Circuit Switched Fall Back and Voice over LTE via Generic Access. The paper analyzes the pros and cons for these solutions and concludes with a suggestion of the best suitable solution for the operators to provide a seamless voice over LTE based on their infrastructure. Also, this paper models the long term solution and proposes an analytical model for simulating the performance of Voice over LTE (VoLTE) that can be used to simulate the behavior of such solution.

Keywords: LTE, SRVCC, VoLGA, CSFB.

1 Introduction

Traditionally, in the second and third generation (2&3G) mobile networks the data has been provided through the Packet Switched (PS) domain and the voice through the Circuit Switched (CS) domain. As an evolution of the 3GPP

Journal of Software Networking, 101–112.
doi: 10.13052/jsn2445-9739.2016.006

networks e.g 2/3G the LTE-4G come to live providing a high data rates up to 100 Mbps (downlink) and 50 Mbps (uplink). The emphasis in LTE has been given to the PS domain since the growth of the data applications on mobile and their greedy need of the bandwidth and speed; as a result the LTE has only PS domain support.

The absence of the CS domain in LTE raise the issue of providing the voice call over LTE which consist of 80% of operators revenue [7] and ensure the continuity with the legacy networks (2/3G).

This paper analyzes the various up to date solutions in the telecommunication market such as the Circuit Switched Fallback [1, 2], Voice over LTE via Generic Access (VoLGA) [3], the Single Radio Voice Call Continuity (SRVCC) [4, 5] and Third Party Voice Over IP [6].

The paper presents and compares these solutions based on the impact on the network infrastructure, the solution complexity and the scalability of the solution to help the operators in eventual decision situation.

The rest of the paper is structured as follow: In Section 2, we review widely the existing solutions such as Circuit Switched Fallback), the voice over LTE via the Generic Access, the soft solution and the 3GPP target and long term solution of providing single voice call continuity. In Section 3, we compare all the solutions while giving their pro and cons. Then and in the last section, we focus on the long term solution SRVCC and propose an analytical model that represents the handover delay interruption for simulating the performance of SRVCC.

2 Voice over LTE Short Term Solutions

2.1 Circuit Switched FallBack (CSFB)

The Circuit Switch Fallback is a short term solution that relies on the existing 2G/3G network. When a User Equipment (UE), camped on LTE, wants to make a CS call, it activates an Extended Service Request (ESR) to the Mobile Management Entity (MME). After receiving the ESR, the MME order the eNodeB to redirect the call to the legacy network 2G/3G. The UE receives the LTE Release Connection and camp on the legacy 2G/3G network and establishes the standard 2G/3G RRC Connection. The downside of this solution is the introduction of the delay due to the procedure execution. This delay varies based on the mobile originating the call, in this case the delay is around 4.7 s [1], or terminating the call, in this case the delay is around 2.84 s [1].

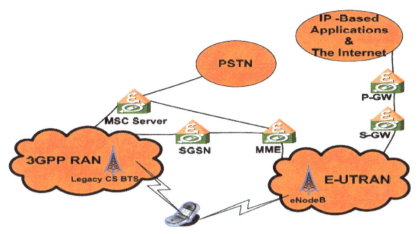

Figure 1   CSFB network architecture.

Furthermore, the CSFB requires an introduction to new interfaces between LTE and legacy 2G/3G particularly the MME-MSC/SGSN and S-GW-SGSN [1] in order to perform the combined registration and paging process.

The benefit of using the CSFB is that the operators can use their existing 2/3G infrastructure to provide voice call.

2.2 The Voice over LTE via Generic Acess (VoLGA)

The Voice over LTE via Generic Access relies on the well-known Generic Access Network (GAN), which consists of adding a Wi-Fi radio interface to the UE becoming dual mode devices that can connect to Wi-Fi over a 2/3G network, when it is available, and connects to the network operator through a GAN gateway. Similarly, the VoLGA use the LTE as a replacement of the Wi-Fi with an introduction of one key critical entity named VoLGA Access Network Controller (VANC) which is a mix between an IP based node and a base station controller node from the LTE and GSM/UMTS point of view respectively. The VANC uses the same interfaces as 2G, 3G and 4G to connect with the mentioned technologies as per the Figure 2.

At the startup of the mobile, the UE register to the LTE and then establishes a secure IPSec tunnel connection to the VANC over LTE [3]. In addition, the UE register to the legacy network through the VANC secure connection that has just been established. The translation between the PS LTE and CS legacy is performed into the VANC entity and is completely transparent to the user and the network.

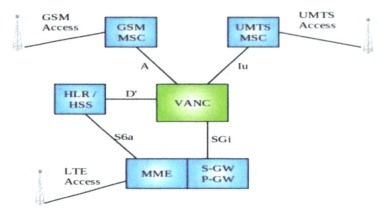

Figure 2   VoLGA network architecture.

Unlike the CSFB, the VoLGA doesn't impact the network entities MME, SGSN and MSC and doesn't increase the call setup procedure. Furthermore, VoLGA provides simultaneously the voice and data services to the UE. However, VoLGA has not been adopted by 3GPP and still requires all visited networks to support VoLGA in order to provide the roaming capability.

2.3 Third Party Software for VoLTE

Another solution that permits voice over LTE is enabled by using a third party software such as Viber or Skype, for providing calls over the LTE as presented in the Figure 3. No changes are required in the network [6], which makes the solution the least expensive among others VoLTE solutions. The downside of

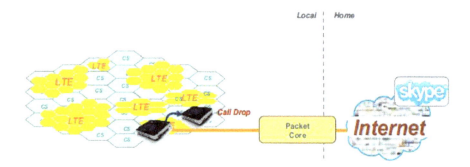

Figure 3   Voice over LTE via skype.

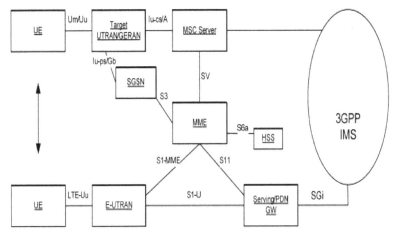

Figure 4   SRVCC architecture for E-UTRAN to UTRAN/GERAN.

this solution is that there is no quality of service guaranteed and this solution doesn't support the CS calls.

2.4 Single Radio Voice Call Continuity

The Single Radio Voice Call Continuity (SRVCC) is the 3GPP target solution, based on IMS, for ensuring the voice continuity between LTE and 2/3G networks. The SRVCC requires that the UE initiates the call using IMS and the Application Server that needs to be implemented in the IMS [4].

The SRVCC uses a combination of the IMS session continuity procedure with a handover procedure. In detail, when a UE reaches the border of the LTE coverage, the E-UTRAN will send a SRVCC handover request to the MME, the latest will trigger the SRVCC procedure with the enhanced MSC (eMSC). The eMSC will perform the session transfer to the IMS and coordinates it with the CS handover to the target cell. Finally the eMSC will send the Forward Relocation Response to MME, which includes the handover command. The Figures 4 and 5 presents the architecture and call flow of the SRVCC [4, 5].

3 Solutions Summary and Comparaison

Deploying a solution relies on the operators' current infrastructure and its future plans. Some operators choose the CSFB because of their existing legacy network infrastructure in place (2/3G), other adopt the VoLGA because the

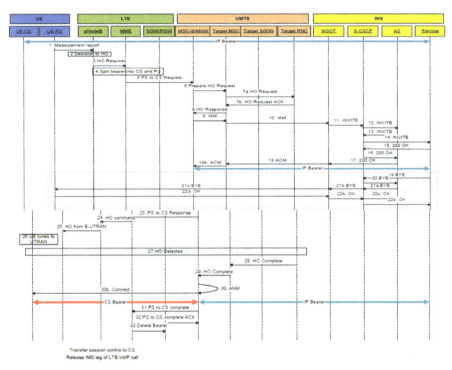

Figure 5   SRVCC call flow from E-UTRAN to UTRAN/GERAN.

availability of such a technology in their network. On the other hand and for a simple and costless, the third party solution will be chosen. For a durable, high and rich multimedia service, the solution based on IMS will be a suitable choice for the operator.

The Table 1 shows a comparison summary of the solutions:

Table 1   Voice over LTE scenarios

	SRVCC	VoLGA	CSFB	Third Party
Availability of 2/3G and neither GAN nor IMS are available	No	Yes, however updates are required at UEs and an VANC need to be introduced	Yes, however the data portion will be slightly affected	Yes, however no QoS is guarantee
2/3G available and plans for IMS	Yes as long term solution	Yes as an alternative solution	Yes as a short term solution	Yes as an alternative solution with no QoS

Table 1 Continued

No 2/3G available and plans for IMS	Yes as long term solution	No	No	No
No 2/3G and no IMS deployment plan	No	No	No	Yes with no QoS
Availability of GAN	Combination of VoLGA and SRVCC for voice service continuity	No		Yes

4 Modeling SRVCC

Many papers have studied the SRVCC handover and tried to come out with a standard model to simulate the performance of this long term solution for voice over LTE.

One of the mathematical models that have been developed is the model presented in [8], in which an analytical expression of the overall delay experienced by the call executing the handover is been presented. This delay was based on the sum of the delay caused by the radio link (RLC) and the delay caused by the remote and internet queuing which was assumed to be following the First In First Out queuing algorithm.

The results of [8] was more related to the radio link delay and has shown that the SRVCC handover interruption time will be reduce under a given Block Error Rate (BLER) with an increased data rate as presented in the Figure 6.

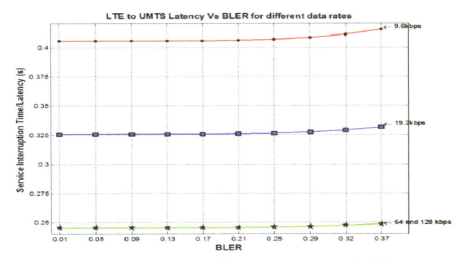

Figure 6 LTE to UMTS SRVCC service interruption time Vs BLER.

In this paper, we will reevaluate the model while assuming that the Quality of Service (QoS) is applied in the whole network since it is crucial for LTE to implement the QoS in order to serve fairly the different traffics that flow through the network.

We estimate that applying a more advanced queuing algorithm will minimize the delay of the SRVCC packets in the network nodes and therefore reduce considerably the overall delay experienced by SRVCC handover which will enhance the performance of the handover.

In order to evaluate the performance of the SRVCC under priority queuing, it is necessary to analyze the interruption time experienced by the ongoing call. This delay, and based on the [9], can be split into three parts: Radio link delay, the network queuing delay and the remote network queuing delay. These delays can be represented by be mathematical models since they are following a unique mathematical behavior.

4.1 The Radio Link Delay

Based on several papers, the radio delay can be measured by analyzing the Radio Link Control (RLC). Assuming the same RLC model in LTE as UMTS, we can represent the delay in the radio part by the following Equation 1 [8]:

$$T_{RLC} = T_{Iub} + (k - 1)\, TTI + \frac{k\,(P_s - (1 - p))}{P_s^2}$$

$$* \left\{ \sum_{j}^{n} \sum_{i}^{j} \left[P\,(C_{ij}) \left(2jT_{Iub} + \left(\frac{j\,(j + 1)}{2} + i \right) * TTI \right) \right] \right\}$$

Equation 1   Radio Delay.

Where:
k: number of frames to be transmitted
n: number of RLC retransmissions
Ps: probability of receiving RLC frame successfully after n transmission
P: probability of RLC frame received erroneously
TIub: latency of the Iub interface
TTI: transmission time interval at eNodeB
P(Cij): the first correctly received frame at destination

4.2 The Network Node and Remote Queuing Delay

The delay caused by the nodes queues relies heavily on the algorithm applied in the system. Many algorithms has been developed and globally used such as FIFO, Priority Queuing (PQ) and Weight Round Robin PQ.

In this research and since LTE is an ALL-IP network that must apply queuing algorithm that guarantee the Quality of Service (QoS) in the network in delivering the traffic, we, first, assume that PQ algorithm has been applied in order to priorize packet coming from the handover process and therefore reducing their treatment delay in the queue.

In order to estimate the average waiting time (delay) in the queues for both the network queuing delay and for the remote queuing delay, we assume that the packets priority is p and arriving with a Poisson distribution with parameter λ_p and their length has an exponential distribution with an average service time $\overline{x_p}$.

We can then calculate the average waiting time W_p of priority p packets in the following way [10]:

$$W_p = \frac{W_0}{(1 - \sigma_p)(1 - \sigma_{p+1})}$$

Equation 2 Average waiting time for PQ.

Where W_0 is the average delay caused by a packet, which is already serviced and calculated as in:

$$W_0 = \sum_{i=1}^{P} \frac{\lambda_i \overline{x^2}}{2}$$

Where $\overline{x^2}$ is the second moment of service time.

And

$$\sigma_p = \sum_{i=p}^{P} \rho_i = \sum_{i=p}^{P} \lambda_i \overline{x_i}$$

After identifying the delay in each part of the involved part of the network, we will deduct the expression of the whole interruption delay experienced by a user that executes the SRVCC handover.

Based on the signal flow in the SRVCC in the Figure 5, we can notice that the service interruption start when the call is released in order to establish a new connection with the 3G network. The Figure 7 [8] represent a zoomed part of the signal flow during the interruption of the service.

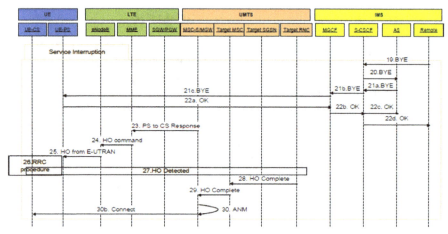

Figure 7   SRVCC service interruption.

Based on the figure and message flow in each node and radio link that is represented by each single arrow, we can write the equation that represent the overall interruption delay as follow:

$$
\begin{aligned}
Dealy_{SRVCC} = \ & Delay_{Remote} + Delay_{Internet} + 2 * Delay_{AS} \\
& + 3Delay_{CSCF} + 2Delay_{MGCF} + Delay_{UE} \\
& + 3Delay_{MSC} + Delay_{MME} + Delay_{eNodeB} \\
& + Delay_{RRC} + Dleay_{RNC}
\end{aligned}
$$

Equation 3   Overall SRVCC interruption time.

Each value of the Equation 3 will be replaced by its expression that has already been presented in the Sections 1 and 2. The Equation 3 will be used to estimate the delay of the interruption of service of the ongoing call until the SRVCC handover has reestablish the circuit with the new cell. The delay will be evaluated under different parameters values in order to study the behavior of the SRVCC performance.

The simulation results and performances evaluation for different priority queuing will be presented in separate paper in the near future.

5 Conclusion

Providing voice calls in LTE has been a challenge and a necessity for the operators since LTE is designed for only PS and the significant incomes generated by the voice calls. Many temporary solutions has been proposed and used, however the operators still seeking for implementing the long term solution SRVCC that will enable the transmission of voice over LTE and supporting the CS over LTE. This solution has not been studied in depth in order to evaluate it performance. This paper has presented the different temporarily solution that are up to date and has focused on the long term solution while proposing a new analytical model that could be used to evaluate the performance of SRVCC. This model will be simulated via Matlab and that for different priority queuing in order to evaluate the performance of SRVCC under these conditions and will be a subject of another future publication.

References

[1] Vargas Bautista, J. E., Sawhney, S., Shukair, M., Singh, I., Govindaraju, V. K., and Sarkar, S. (2013). Performance of Fallback from LTE to UMTS Qualcomm Corporate Engineering. *IEEE Commun Mag.* 51, 136–143.
[2] CS fallback in Evolved Packet System. (2010). Technical Specification 3GPP TS 23.272.
[3] Voice over LTE with Generic Access. (2010). Technical Specification VoLGA Stage 2.
[4] Single Radio Voice Call Continuity. (2008). Technical Specification 3GPP TS 23.216.
[5] Single Radio Voice Call Continuity. (2010). Technical Specification 3GPP TR 23.856.
[6] Gavrilovic, S. (2010). "Standard Based Solutions for Voice SMS Services over LTE. *IEEE/MIPRO* 334–339.
[7] Cox, C. (2012). *Introduction to LTE*. Hoboken, NJ: Wiley.
[8] Namakoye, J., Van Olst, R. (2011). A Performance evaluation of a voice call handover scheme between LTE and UMTS. Rome: IEEE Conference Publications.
[9] Kim, K. (2009) "A seamless voice call handover scheme for next generation cellular network," in *IEEE 15th Asia-Pacific Conference on Communications* (Shanghai: IEEE).
[10] Balogh, T., and Medvecki, M. (2010). Comparison of Priority Queuing Based Scheduling Algorithms. *ElektroRevue*.

Biographies

A. A. El Arby is a Ph.D. student at the University of Gaston Berger, Senegal since 2012. A Ph.D. that has been started at the Ecole de Technologie Superieure, Canada where he received his M.Sc. and Engineering diploma in telecommunication. He is Professor in Networking and Telecommunication, in Nouakchott University and ISCAE Mauritania. Author of published papers in several journals such as IEEE and Springer.

O. Thiare, received a Ph.D. in computer science (Distributed systems) at 2007 from the University of Cergy Pontoise, France. He is Full Professor in Computer science, in Gaston Berger University of Saint-Louis Senegal. He has been author and co-author of published papers in several journals and recognized international conferences and symposiums.

Implementation of Noise Reduction Methods for Rear-View Monitoring Wearable Devices

Huy Toan Nguyen[1], Seung You Na[1,*], Jin Young Kim[1]
and Gwon Seok Sun[2]

[1]School of Electronics and Computer Engineering, Chonnam National University, Gwangju, Republic of Korea
[2]Korea Polytechnics College, Gwangju, Republic of Korea
E-mail: nguyenhuytoantn@gmail.com; beyondi@jnu.ac.kr; mtrsks@hanmail.net
*Corresponding Author: syna@jnu.ac.kr

Received 15 September 2016; Accepted 15 November 2016;
Publication 24 November 2016

Abstract

This study suggests effective noise reduction methods for wearable neckband devices, which are able to monitor users' rear-view areas. The wearable neckband device helps the user to monitor rear-view areas in which he/she is unable to see in normal ways (without turning back). Unlike general computers or supercomputer systems, the neckband devices have some particular constraints such as small size, lightweight and low power consumption. In a general vision system, there are many kinds of noises, which significantly decrease system quality such as impulse noise, random noise, motion noise, etc. These noises also affect wearable devices, which use cameras as the system input. Moreover, when the user walks or runs, the neckband device moves accordingly. The changing position of the neckband device causes many other noises such as camera motion (ego-motion) noises. Furthermore, when the user walks from indoors to outdoors or vice versa, the illumination dramatically changes, which also affects the device performance. Effective noise reduction methods to deal with these noises are proposed in this study. Random noise and other small noises are removed by using a Gaussian filter

Journal of Software Networking, 113–136.
doi: 10.13052/jsn2445-9739.2016.007

and adaptive color threshold techniques. We propose to use feature detection and the homography matrix estimation method to reduce ego-motion noise. Remaining noises are cancelled out by a morphology technique. Finally, we apply Local Binary Patterns (LBP) descriptor and Adaboost classifier to classify whether there are people or not in the moving foreground object regions. The experiments demonstrate that our proposed noise reduction methods have achieved successful results in the different environments and users' walking speeds.

Keywords: Noise reduction, Rear-view monitoring, Wearable device.

1 Introduction

Surveillance systems are used to monitor potentially dangerous objects in region of interest [1]. A wide range of applications have been buttressed upon the advances in monitoring techniques such as search and rescue [7], video surveillance [8], vehicle driving assistance [9], reconnaissance [10] and robotics [11]. The reliability and robustness of such systems strictly depend on the environmental conditions, for instance, light intensity. There have been many studies attempting to eliminate the unwanted effects of unpredictable incoming light sources [14, 15, 24–26].

Nowadays, with the advent of semiconductor and electronic technologies, electronic components are becoming more and more affordable. Thanks to the development of digital cameras, video monitoring systems based on computer vision are becoming popular in daily life. Due to that reason, various approaches to building autonomous systems based on image processing have been published [12, 13]. Their works were based on computers or supercomputers to achieve high accuracy and reduce processing time. We propose the design and implementation of a small wearable device to detect and classify dynamic objects in a rear-view area of the user based on microprocessors.

Developing a small wearable device with high accuracy and short processing time is a challenge due to low resolution, background changes, and hardware constraints. In case of wearable devices, there are strong variations in illumination, background, shadows and other random types of noise such as impulse noise [14], Gaussian noise, Poisson noise [15], Speckle noise, Salt and Pepper noise, which can make the situation even more complicated. To address this problem, various effective noise reduction methods for small wearable devices are studied and exploited in this paper.

2 Related Work

2.1 Video Monitoring System

There has been increasing interest in building monitoring systems. However, a video monitoring system has many constraints such as low processing time, high accuracy, affordability and ease for the user. To address these problems, a lot of research work has been proposed. O. Jafari *et al.* [2] and D. Mitzel *et al.* [3] suggested a system, which used a single CPU core. They used a head-mounted camera based on Kinect RGB-D input data to detect pedestrians in a close-range and achieved a high processing speed up to 18 fps. An application using IP cameras to send and receive data via network and Internet was proposed by R. Rashmi *et al.* [4]. They used Motion Detector application to warn the user via email or text messenger. The image data were transferred from mobile to PC to store data for a long time. Lefang *et al.* [5] came up with a remote video monitoring system based on ARM processing chip using Linux operating system combined with GPRS technologies. X. Jiangsheng [6] introduced a video monitoring system based on TMS320DM642 DSP, local network and GPRS for large railway maintenance machinery. However, these previously mentioned systems were mainly based on computers or supercomputers for processing acquisition videos. The high cost and inconvenience for a single user limit the popularity of these systems in real life. The design and implementation of a wearable device, which is small and lightweight as SenseCam device [16] with real-time processing, are currently hot topics in the computer vision and robotics communities.

2.2 Wearable Device

The design and development of wearable devices have accumulated lots of attention from the scientific community and industry for the last few years [17, 18]. There are various applications based on wearable devices, especially for health care problems. In [19], A. Pantelopoulos *et al.* designed and developed wearable biosensor systems for health monitoring. The wearable biosensor system included many parts such as physiological sensors, transmission modules and processing capabilities for health monitoring. An overview of state-of-art wearable technologies for remote patient monitoring was presented by Hung *et al.* [20]. Their research work suggested developing tele-home healthcare systems, which used wearable devices for monitoring remotely. J. Hernandez *et al.* [21] proposed to use sensors embedded in the Google Glass. In [22], C. Setz *et al.* introduced the wearable device for detecting early warning signs at the workplace. They took interest in developing a method to analyze the discriminative power of electrodermal activity (EDA) to detect stress signals. On the other side, wearable devices were adopted to make navigation systems to assist impaired people. In [23], a survey about wearable devices, which were used for obstacles avoidance purposes, was presented. The author introduced many wearable devices for assisting impaired people. Three main categories of wearable navigation systems were electronic travel aids (ETAs), electronic orientation aids (EOAs), and position locator devices (PLDs). Most of the approaches introduced so far consider wearable devices for healthcare and obstacles avoidance purposes with active sensors such as distance sensors or other sensors for gathering input signals. Due to the limitation of active sensors, wearable monitoring systems based vision sensors have been increasing. However, only few approaches provided experimental results for monitoring at human blind spots based on vision sensors. In this paper, we develop noise reduction methods, which can be applied to the small wearable device based on microprocessor using camera as an input sensor.

2.3 Noise Reduction Method

The output quality of a wearable neckband device with a camera can be significantly decreased by various components of noise. There are some sources of noise such as camera sensor, random noise, and motion noise. Recently, there has been growing interest in noise reduction and enhancement of the output quality. In [24], M. Kim *et al.* proposed an approach to reduce noise and enhance quality of extremely low-light video. They used adaptive temporal filtering based on Kalman filter to reduce noise. Another method for denoising color video sequences was introduced by V. I. Ponomaryov *et al.* in [25]. They selected fuzzy filtering to obtain better results. The image denoising method based on local average gray values and gradients, pre-classifying neighbourhoods and thereby reducing the original quadratic complexity to a linear one and the influence of less-related areas in the denoising of a given pixel was presented by M. Mahmoudi *et al.* in [26].

Different from the aforementioned methods, we propose a noise reduction method to eliminate both static and motion noise for rear-view monitoring wearable devices. We rely on Gaussian Filter and morphology technique to overcome static noise. The ego-motion noise caused by the camera variation is compensated by feature based and homograph decomposition method.

We propose adaptive color threshold method and apply Adaboost classification using LBP features to achieve useful information from input captured video.

We organize our paper as follows. In Section 3, the system architecture including system hardware and system software is first explained. The experiment setup is fruitfully represented in Section 4. Results and discussions are in Section 5. Finally, conclusion and future works are mentioned in Section 6.

3 System Architecture

Unlike a general computer or supercomputer, which can execute multiple tasks, the proposed wearable neckband device system only performs monitoring tasks with the specific hardware constraints. The system hardware includes a set of constraints such as small sizes, light weight, low costs, and high processing ability. The proposed algorithms for noise reduction have been designed and implemented by the system software. In the next parts, the system hardware and software are discussed in detail.

3.1 The System Hardware

In general, the system performance depends on the system hardware. The proposed system hardware is divided into several parts. The schematic diagram is presented in Figure 1.

3.1.1 Camera module
The camera module is one of the most important and vital elements in any vision system. The camera's quality can directly affect the system performance. Due to the system constraints, a small, inexpensive, and acceptable

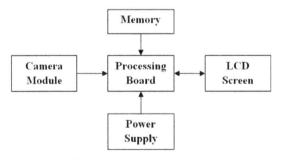

Figure 1 Hardware system.

quality camera is selected. Captured images are transmitted to the processing board via a lightweight cable.

3.1.2 Processing board

A small, lightweight and user-friendly Processing Board (PB) is selected. The CPU on PB takes the video data input from the camera module, processes all logical and computational tasks and gives the output video result on the LCD monitor. In our case, CPU and memory are integrated on the same PB. In the selection of PB, low power consumption for the long time operation is considered as one of vital properties.

3.1.3 Memory

Another basic element of system hardware is the memory. The memory sizes can significantly affect the system performance and time consumption. In the general case, the memory is separated into two parts: Random Access Memory (RAM) and storage memory. All the temporary data and variables are stored in the RAM, while the system is turned on. Another memory is used for storing the operating system and software algorithms. All memories are physically integrated into PB.

3.1.4 Power supply

When designing small devices such as wearable devices, low power consumption is highly considered. For the long time operation, the proposed system components are selected to reduce power consumption. The system software is also optimized for the same purpose. For convenient and long time operation, a small size, high output capacity and suitable shape battery is selected.

3.1.5 LCD screen

After processing steps, the results are displayed on the LCD screen. To make the system more user-friendly and interactive, the touch screen LCD is installed. The display screen not only provides certain information to the user but also works as the interface between the user and the system device. All interaction tasks are performed on the touch LCD screen. The LCD screen, which is almost the same size as the PB, is chosen to decrease the device's physical size.

3.2 The System Software

The system software describes all performing logical and mathematical calculations on the input video data. The output of the system is presented on the LCD screen. With the specific system hardware, the good system software tends to achieve real-time and high accuracy with low power consumption. In this section, an algorithm for monitoring a moving object in a rear-view range is proposed.

The input image data is captured from the mounted camera on the wearable neckband device when the user walks. The distance from the user to objects is in a close-range of 1 to 10 meters. The proposed method has three main steps: pre-processing, detection of moving foreground objects and object classification. The methods of reducing noise in a captured image are proposed in the pre-processing step. Noises, which mainly come from the illumination changes and the camera sensor, are reduced in this step. The primary purpose of the second step is to detect moving foreground objects. Moreover, in the second step, methods of decreasing camera motion noises and the computation time are considered. Finally, object feature extraction and object classification within the moving object regions are introduced in the third section. The flowchart of the system software is presented in Figure 2.

3.2.1 Pre-processing step

The input captured image is distorted by many kinds of noise such as impulse noise, random noise, Gaussian noise and so on. The quality of the system is directly affected by these noises. Due to hardware limitations and real-time constraints, a simple filter is selected to remove these noises. At the beginning, when the camera module is activated, the input captured image $I(x, y, t)$ immediately is resized to 320×240 pixels and converted to grayscale $I^{gray}(x, y, t)$. The Gaussian Filter is adopted for noise reduction on the converted grayscale image. The Gaussian Filter can be presented by Equation (1):

$$G_0(x, y) = I e^{\frac{-(x-\mu_x)^2}{2\sigma_x^2} + \frac{-(y-\mu_y)^2}{2\sigma_y^2}} \tag{1}$$

where μ is the mean and σ represents the variance (for each of the variables x and y).

We selected the kernel size equal to 5×5. After applying Gaussian filter, small noises such as salt and pepper noise and random noise are cancelled out.

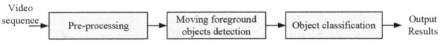

Figure 2 Software system.

3.2.2 Moving foreground objects detection

The camera is mounted on the neckband for detection task. The main reason of background instability is the moving action of the user when he/she uses the wearable system. The camera ego-motion and the motion of moving objects are considered as two fundamental motions in the detection problem. In case of non-static background, the camera motion has to be estimated to know which parts of the picture are changed due to the camera motion and the other parts of the picture are changed independently. In [27], D. Szolgay *et al.* adopted Hierachical Block-Matching algorithm to estimate the strong camera motion. The main problem with Hierachical Block-Matching algorithm is the computation cost. In this paper, we propose a method to overcome the weakness of the above method.

Let's consider two grayscale consecutive images $I^{gray}(x, y, t-1)$ and $I^{gray}(x, y, t)$. In the ideal case, the relationship between two grayscale consecutive images can be presented by Equation (2) or Equation (3):

$$I^{gray}(x, y, t) = H I^{gray}(x, y, t-1) \tag{2}$$

or

$$\begin{pmatrix} x_t \\ y_t \\ 1 \end{pmatrix} = \begin{bmatrix} h_{11} & h_{12} & h_{13} \\ h_{21} & h_{22} & h_{23} \\ h_{31} & h_{32} & h_{33} \end{bmatrix} \begin{pmatrix} x_{t-1} \\ y_{t-1} \\ 1 \end{pmatrix} \tag{3}$$

where H is a homography matrix.

It is necessary to extract moving foreground objects in order to reduce calculation time for object classification step. Through experiments with different features, Good Feature to Track [28] is suitable for our system in comparison with other methods [29, 30, 31, and 32]. In order to track key points from the previous grayscale image $I^{gray}(x, y, t-1)$ to the current grayscale image $I^{gray}(x, y, t)$, we propose to use the tracking method [33]. The projective transformation matrix between the two consecutive images, a homography matrix H, is determined by the RANSAC method [34]. From the homography matrix, we figure out the reference motion-compensated image by Equation (4).

$$I^{gray}_{Ref}(x, y, t) = H I^{gray}(x, y, t-1) \tag{4}$$

Error image $E^{gray}(x, y, t)$, which is determined by subtraction of $I^{gray}(x, y, t)$ from $I^{gray}_{Ref}(x, y, t-1)$, presents the differences between an ideal image result and a real image result. The error image $E^{gray}(x, y, t)$

presents moving foreground objects independently from the camera motion as follows:

$$E^{gray}(x, y, t) = \left| I_{\text{Ref}}^{gray}(x, y, t-1) - I^{gray}(x, y, t) \right| \tag{5}$$

In an ideal situation, the error image $E^{gray}(x, y, t)$ consists of only moving foreground objects. However, in our real experiments, the camera motion, in some cases, is even stronger than movements of an object itself. As a result, the error image also contains false positive pixels, which significantly degrade the system quality. Using normal methods, which are based on gray scale images to get moving foreground objects, is impossible in this case. Thus, we propose to use color information from the original image $I(x, y, t)$. The proposed error image is presented by the following equation:

$$E(x, y, t) = \begin{cases} 0 \text{ if } E^{gray}(x, y, t) = I^{gray}(x, y, t) \text{ or } E^{gray}(x, y, t) < th_E \\ I(x, y, t) \text{ otherwise} \end{cases} \tag{6}$$

Finally, the threshold error image th_E is calculated by the mean and variance of error image in each R, G, and B channel separately. This threshold is adaptive for each specific error image. Finally, we apply a morphology method to remove small noise and get moving foreground object regions.

3.2.3 Object classification

After the pre-processing step, the random noise, illumination noise and other motions noises should have been removed already in the video dataset. The output of Section 2 is regions of interest (ROI), which include only moving foreground objects. In order to get more detail information, we propose the object classification step to determine whether moving objects are human or non-human.

In our experiments, moving objects are in a rear-view range in the distance of 1 to 10 meters. We realize that when the user wears the neckband device, in a rear-view and close range, the camera is unable to capture the full-body images. Due to this reason, the full-object detector methods [35, 36] are not suitable for our cases. In this study, for the object classification purpose, we only concentrate on human face detection in a rear-view area and close range distance.

Feature selection and classification have demonstrated promising results for this task. Among many feature descriptors for the object classification, we propose to use a Local Binary Pattern descriptor and an AdaBoost classifier. The original LBP [37] forms labels for the image pixels by using the threshold

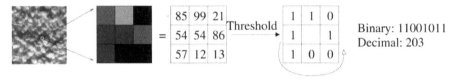

Figure 3   The basic LBP operator [30].

of the 3×3 neighborhood of each pixel with the center value and then by considering the result as a binary number.

The basic methodology for LBP based on face description was proposed by Ahonen *et al.* [38]. The facial image is divided into local regions, and LBP texture descriptors are extracted from each region independently. The basic LBP is shown in Figure 3.

Due to the necessity of the fast response, we select an Adaboost algorithm [39] for classification tasks.

4 Experiments

4.1 Hardware Selection

Component selection is one of the most challenging tasks for designers or researchers, in which they have to have a balance between two conflicting requirements, which are low cost and high performance.

In our experiment, due to some constraints such as small in size, high processing ability, low-power consumption, and lightweight, Raspberry Pi 2 model B (RPI) is selected as the main processing board. All system software steps are implemented on this processing board. Raspberry Pi 2 model B is a credit card sized single-board computer. With a small size, this processing board is suitable for the single user to use our product.

To get the high quality input video data, Raspberry Pi camera module is selected as the system input. Raspberry Pi camera module has a 5-Megapixel camera, which can record high-definition (HD) videos up to 720p resolution, with high sensitivity, low noise and low crosstalk features. The maximum capture rate is up to 60 frames per second (fps). Moreover, Raspberry Pi camera module has extremely small size ($25 \times 20 \times 9$ mm) and lightweight (3 grams), which is sufficient to make wearable devices as neckband devices.

Furthermore, the RPI has 1 GB RAM, which is enough for real-time processing. The 16 GB SDCard is selected as the external storage of the system.

The operation system and all software are stored in the external storage. The reading and writing speed of SD card also affect the system performance.

Results are presented on the LCD screen. In our experiment, we choose the LCD touch screen. The LCD screen is the same size as the PB. The touch LCD screen not only provides useful information, but can also work as the interface for the device.

Finally, the battery is selected for our system. We choose a small size, high capacity, and suitable shape battery for a wearable device. The proposed neckband device hardware is shown in Figure 4.

4.2 Datasets

To evaluate proposed noise reduction methods, datasets are recorded in real conditions where the illumination changes and the user keeps on walking while the system records video. The datasets are recorded in different environments including various indoor and outdoor environments. The user's walking speed is considered as one of the key elements, which affects the final results. The user's walking speed is classified into three categories: slow, normal, and fast speed. The details about datasets are provided in Table 1.

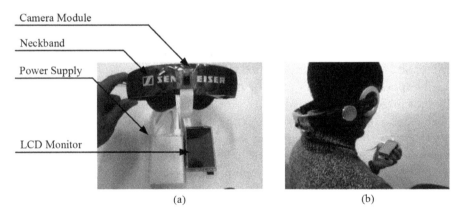

Figure 4 The neckband device. (a) Hardware structure, (b) Setup device for a user.

Table 1 Test dataset description

Video Sequence	Environments	Walking Speed	Frames
Video 1	Indoor	Fast	166
Video 2	Indoor	Normal	262
Video 3	Indoor	Slow	283
Video 4	Outdoor	Slow	286

5 Results and Discussions

The software algorithms are implemented by using the OpenCV library based on Python programming language. To evaluate the running time of our algorithm on proposed platform, we compute the average running time on each step by the following Equation:

$$\bar{t} = \frac{\sum\limits_{i=1}^{N} t_i}{N} \tag{7}$$

where N is the number of frames in one video sequence, t_i is processing time for frame i, $\bar{t}$ is the average time consumption for whole video. It is computed separately for each step for each video. Finally, we compute the average consumption time for four videos by Equation (7) with N is the video number, and t_i is the processing time for video i. The average processing time for each step is presented in Table 2.

Based on Table 2, the average total running time for one frame, which has frame size of 320×240 pixels, is approximately equal to 771 milliseconds. The Pre-processing step for reducing small static noise by Gaussian Filter takes around 11 milliseconds to complete. Moving foreground object detection takes the longest time due to a number of sub-steps inside such as feature detection, feature tracking, homography matrix estimation and adaptive color threshold technique. The human face detection in region of interest (ROI) based on LBP descriptor and Adaboost classify method takes 200 milliseconds. This method is mainly applied when users walk or run, so it is possible to be real-time method for these applications.

We extend our previous work [1] by comparing feature extraction processing times on one frame with different techniques. The approximate running time per frame for each method is computed by Equation (7). Moreover, the average number of detected key points in each frame is estimated in Equation (8):

Table 2   Average processing time for each step per frame

Step	Time Consumption (second)
Pre-processing	0.0109
Moving foreground object detection	0.5634
Face classification	0.1969
Total Time	0.7712

$$\bar{k} = \frac{\sum\limits_{i=1}^{N} k_i}{N} \qquad (8)$$

where N is the number of frame in one video sequence, k is number of detected key points in frame i^{th}, $\bar{k}$ is the average feature detected in video sequences. The detailed results and samples are presented in Table 3 and Figure 5.

Table 3 Average runtime feature extractions per frame

Features	Number of Key Points	Time Consumption (seconds)
Good Feature to Track [28]	52	0.0147
FAST [29, 30]	479	0.0088
ORB [31]	446	0.0315
BRIEF [32]	55	0.0128

(a)   (b)

(c)   (d)

Figure 5 Comparison of features detection methods. (a) Good Features to Track, (b) FAST features, (c) ORB features, (d) BRIEF features.

We realize that various other feature detection methods such as FAST [29, 30] and BRIEF [32] consume less time to detect feature points in compare with Good Feature to Track [28]. However, after several experiments, we find out that other feature detection methods such as FAST, ORB and BRIEF produce higher wrong detection points for the images with moving objects. Since homography matrix is calculated based on detected feature points, wrong detected points lead to fail in calculation of homography matrix. In order to achieve high accuracy and acceptable processing time, we choose Good Feature to Track method for detecting key points in our experiments.

Recently, in [40, 41], authors proposed the process execution of object classification based on Haar-features with the same platform. We adopt LBP descriptors and Adaboost classification, which is distinct from previous published studies to achieve better running time. The results are shown in Table 4.

Average processing time per frame is calculated and compared with other image denoising and object detection methods as presented in Table 5.

The accuracy of our proposed methods for noise cancelling due to the user's walking movements is presented in Table 6. In this part, we test our algorithms on several video sequences with different walking speeds and various environments.

Table 6 show that Video 3, which is recorded in indoor environments with the slow walking speed, give the best accuracy result. When the user walks faster, the accuracy decreases due to calculation error of homography matrix.

Table 4   Comparison of face classification performance

Features	Time Consumption (second)
Haar-features [40, 41]	0.556
LBP descriptors	0.1969

Table 5   Average processing time per frame

Method(s)	Application	Processing Time (seconds)
M. Kim *et al.*, 2015 [24]	Image denoising	6.8
M. Mahmoudi *et al.*, 2005 [26]	Image denoising	481
D. Szolgay *et al.*, 2011 [27]	Moving object detection	39.4138
R. J. Moreno, 2014 [40]	People detection	2.5
W. F. Abaya *et al.*, 2014 [41]	Security camera at night	2.5
Our method	Noise Reduction & Moving object detection	0.7712

Table 6 The system performance

Video Sequence	Environments	Walking Speed	Accuracy
Video 1	Indoor	Fast	69.71%
Video 2	Indoor	Normal	73.04%
Video 3	Indoor	Slow	88.70%
Video 4	Outdoor	.Slow	81.48%

With the same speed but different environments, the accuracy is different, due to the vibrational background. In indoor environments, we get more stable key points from the non-moving objects such as walls and doors. However, in the outdoor case, the detected key points may not only on a static object but also from some small moving objects such as trees and flags moved by wind.

For better understanding of the proposed method, we illustrate a sample on two consecutive images, called previous image and current image, in our video sequences as shown in Figure 6. First, two color input images are converted to grayscale images as shown in Figure 6(a) and 6(d). Then, Gaussian Filter is applied to remove static noises such as random noise and other small noises. Results of static noise cancellation method are presented in Figure 6(b) and 6(e). The feature points are detected from the previous image as in Figure 6(c). These feature points are tracked from the previous image to current image, respectively. In case of static camera, we can apply simple background subtraction technique to achieve foreground objects. However, it is impossible to achieve only foreground object regions if we apply the same process for an unstable camera as our cases, as shown in Figure 6(g). For that reason, the homography matrix H is calculated based on detected feature points on the previous image and tracked feature points in the current image. After applying Equation (4) and Equation (5) respectively, we get the error image as illustrated in Figure 6(h). Due to the lack of information about foreground objects, we continuously correct the error image by Equation (6), and the results are shown in Figure 6(i). The foreground objects are determined from this modified error image, which is estimated by white region in Figure 6(j). Figure 6(k) illustrates the region of interest (foreground moving objects), which is shown in the green rectangular box. Finally, face classification based on LBP descriptor and Adaboost classifier in the region of interest is shown in Figure 6(l). After a number of steps, the user can know whatever there are moving objects or not in blind areas regardless of human or non-human objects. The results show that our algorithm is able to work well enough to reduce various kinds of noises and correctly classify human or non-human objects.

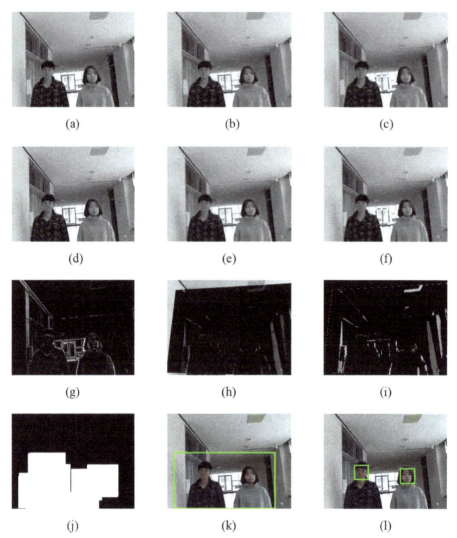

Figure 6   The sample result. (a) (d) Two grayscale consecutive images, (b) (e) Two consecutive images after removing static noises, (c) Detected features in the previous image, (f) KLT tracker features, (g) Normal subtraction, (h) Error image, (i) Modified error image, (j) Region of Interest, (k) Moving foreground objects, (l) Face detection.

6 Conclusions and Future Works

A design and implementation of a wearable neckband device for rear-view monitoring is presented in this study. A monitoring system has to face various types of noise. Some of them are from the hardware electronic devices while others are from the outside environment. Effective noise reduction methods are proposed in this paper to deal with different kinds of noises. We propose combining a Gaussian filter, an adaptive color filter and a homography matrix to reduce the external noise effects. The Gaussian filter is adopted to reduce common noises such as Salt and Pepper noise, which is caused by camera quality. We choose Good Feature to Track method to extract key points and the homography matrix to estimate moving foreground objects. We propose an adaptive color threshold method to get moving foreground objects. The remaining noise is removed by morphology technique. Finally, we apply the LBP descriptor and Adaboost to classify as human or non-human in the video frame.

The proposed noise reduction process for wearable neckband device is able to execute high accuracy when the user walks with a normal speed. The static and motion noises are removed by combining homography decomposition technique and adaptive color method. The proposed method is investigated using a small microprocessor device. It is possible to increase the accuracy of the system in the future. In the next stage, we will extend this research to other platforms in case of camera moving with a higher speed. The above results of noise reduction method can be applied to other real devices for monitoring blind spots around cars, search and rescue based drones or even for mobile navigation.

Acknowledgments

This study was financially supported by Chonnam National University, 2016.

References

[1] Nguyen, H. T., Choi, Y. S., Sun, G. S., Na, S. Y., and Kim, J. Y. (2016). Effective noise reduction methods for rear-view monitoring devices based on microprocessors. *Mobile Wirel. Technol.* 391, 51–58.
[2] Jafari, O., Mitzel, D., and Leibe, B. (2014). "Real-time RGB-D based people detection and tracking for mobile robots and head-worn cameras," in *ICRA International Conference on Robotics and Automation*. RWTH Aachen University, Germany.
[3] Mitzel, D., and Leibe, B. (2012). "Close-range human detection for head-mounted camera," in *BMCV British Machine Vision Conference*. RWTH Aachen University, Germany.
[4] Rashmi, R., and Latha, B. (2013). "Video surveillance system and facility to access Pc from remote areas using smart phone," in *ICICES*

International Conference on Information Communication and Embedded System (Rome: IEEE), 491–495. doi: 10.1109/ICICES.2013.6508393

[5] Lefang, Z., Jian-xin, W., and Kai, Z. (2013). "Design of embedded video monitoring system based on S3C2440," in *ICDMA International Conference on Digital Manufacturing and Automation* (Rome: IEEE), 461–465. doi: 10.1109/ICDMA.2013.108

[6] Jiangsheng, X. (2011). "Video monitoring system for large maintenance machinery," in *ICEMI International Conference on Electronic Measurement and Instrument* (Rome: IEEE) 3, 60–63. doi: 10.1109/ICEMI.2011.6037855

[7] Morse, B. S., and Engh, C. H. (2010). "UAV video coverage quality maps and prioritized indexing for wilderness search and rescue," in *Proceedings of the 5th ACM International Conference on Human-Robot Interaction* (Rome: IEEE), 227–234.

[8] Tseng, B. L., Lin, C. Y., and Smith, J. R. (2002). "Real-time video surveillance for traffic monitoring using virtual line analysis," in *International Conference on Multimedia and Expo* (Rome: IEEE), 2, 541–544.

[9] McCall, J. C., and Trivedi, M. M. (2006). Video-based lane estimation and tracking for driver assistance: survey, system, and evaluation. *IEEE Trans. Intell. Transport. Syst.* 7, 20–37.

[10] Bhaskaranand, M., and Gibson, J. D. (2013). Low Complexity Video Encoding and High Complexity Decoding for UAV Reconnaissance and Surveillance. *Int. Symp. on Multimedia*, 163–170.

[11] Li, Z., Yang, C., C-Y Su, Deng, J., and Zhang, W. (2016). Vision-Based Model Predictive Control for Steering of a Nonholonomic Mobile Robot. *IEEE Trans. Control Syst. Technol.* 24, 553–564.

[12] Arroyo, R., Yebes, J. J., Bergasa, L. M., Daza, I. G., and Almazán, J. (2015). Expert video-surveillance for real-time detection of suspicious behaviours in shopping malls. *Int. J. Expert Syst. Appl.* 42, 7991–8005.

[13] Guler, P., Emekksiz, D., Temizel, A., Mustafa Teke, and Temizel, T. T. (2016). Real-time multi-camera video analytics system on GPU. *J. of Real-time Image Processing* 11, 457–472.

[14] Yadav, P. (2015). "Color image noise removal by modified adaptive threshold median filter for RVIN.," in *EDCAV International Conference on Electronic Design, Computer Networks and Automated Verification* (Rome: IEEE), 175–180.

[15] Foi, A., Trimeche, M., Katkovnik, V., and Egiazanrian, K. (2008). Practical Poissonian–Gaussian Noise Modeling and Fitting for Single-Image Raw-Data. *IEEE Trans. Image Process.* 17, 1737–1754.

[16] Hodges, S., Williams, L., Berry, E., Izadi, S., Srinivasan, J., Butler, A., G. Smyth, Kapur, N., and Wood, K. (2006). "SenseCam: a retrospective memory aid," in *International Conference of Ubiquitous Computing* (Berlin: Springer), 177–193.

[17] Lv, Z., Feng, S., Feng, L., and Li, H. (2015). Extending touch-less interaction on vision based wearable device. *In Virtual Reality (VR)*, 231–232.

[18] Woodberry, E., Browne, G., Hodges, S., Watson, P., Kapur, N., and Woodberry, K. (2015). The use of a wearable camera improves auto-biographical memory in patients with Alzheimer's disease. *J. Memory*, 23, 340–349.

[19] Pantelopoulos, A., and Bourbakis, N. G. (2010). A Survey on Wearable Sensor-Based Systems for Health Monitoring and Prognosis. *IEEE Trans. Syst. Man Cybern C Appl. Rev.* 40, 1–12.

[20] Hung, K., Yang, Y. T., and Tai, B. (2004). Wearable medical devices for tele-home healthcare. *Conf. Proc. IEEE Eng. Med. Biol. Soc.* 7, 5384–5387.

[21] Hernandez, J., Li, Y., Rehg, J. M., and Picard, R. W. (2014). "BioGlass: Physiological parameter estimation using a head-mounted wearable device," in *Proceedings of the 2014 EAI 4th International Conference on Wireless Mobile Communication and Healthcare (MOBIHEALTH 2014)*, 55–58, Athens.

[22] Setz, C., Arnrich, B., Schumm, J., Marca, R. L., Tröster, G., and Ehlert, U. (2010). Discriminating stress from cognitive load using a wearable EDA device. *IEEE Trans. Inf. Technol. Biomed.* 14, 410–417.

[23] Dakopoulos, D., and Bourbakis, N. G. (2010). Wearable obstacle avoid-ance electronic travel aids for blind: a survey. *IEEE Trans. Syst. Man Cybernetics C* 40, 25–35.

[24] Kim, M., Park, D., Han, D. K., and Ko, H. (2015). A novel approach for denoising and enhancement of extremely low-light video. *IEEE Trans. Consum. Electron.* 61, 72–80.

[25] Ponomaryov, V. I., Montenegro-Monroy, H., Gallegos-Funes, F., Pogreb-nyak, O., and Sadovnychiy, S. (2015). Fuzzy color video filtering technique for sequences corrupted by additive Gaussian noise. *J. Neurocomputing* 155, 225–246.

[26] Mahmoudi, M., and Sapiro, G. (2005). Fast image and video denoising via nonlocal means of similar neighborhoods. *IEEE Signal Process. Lett.* 12, 839–842.

[27] Szolgay, D., Benois-Pineau, J., Megret, R., Gaestel, Y., and Dartigues, J. F. (2011). Detection of moving foreground objects in videos with strong camera motion. *J. Pattern Anal. Appl.* 14, 311–328.

[28] Shi, J., and Tomasi, C. (1994). "Good features to track," in *Proceedings of the CVPR International Conference. on Computer Vision and Pattern Recognition*, Seattle, WA, 593–600.

[29] Rosten, E., and Drummond, T. (2006). "Machine learning for high speed corner detection," in *Proceedings of the ECCV the 9th European Conference on Computer Vision*, Graz, 430–443.

[30] Rosten, E., Porter, R., and Drummond, T. (2010). Faster and better: a machine learning approach to corner detection. *IEEE Trans. Pattern Anal. Mach. Intell.* 32, 105–119.

[31] Rublee, E., Rabaud, V., Konolige, K., and Bradski, G. R. (2011). "ORB: an efficient alternative to SIFT or SURF," in *Proceedings of the ICCV International Conference on Computer Vision* (Rome: IEEE), 2564–2571. doi: 10.1109/ICCV.2011.6126544.

[32] Calonder, M., Lepetit, V., Strecha, C., and Fua, P. (2010). "BRIEF: binary robust independent elementary features," in *Proceedings of the ECCV the 11th European Conference on Computer Vision*, Heraklion, 778–792.

[33] Tomasi, C., and Kanade, T. (1991). *Detection and Tracking of Point Features*. Technical Report CMU-CS-91-132. Pittsburgh, PA: Carnegie Mellon University.

[34] Fischler, M., and Bolles, R. (1981). Random sample consensus: a paradigm for model fitting applications to image analysis and automated cartography. *Commun. ACM* 24, 381–395.

[35] Dalal, N., and Triggs, B. (2005). "Histograms of oriented gradients for human detection," in *Proceedings of the CVPR Conf. on Computer Vision and Pattern Recognition* (Washington, DC: IEEE Computer Society), 886–893.

[36] Felzenszwalb, P., Girshick, R., McAllester, D., and Ramanan, D. (2010). Object detection with discriminatively trained part-based models. *IEEE Trans. Pattern Anal. Mach. Intell.* 32, 1627–1645.

[37] Ojala, T., Pietikäinen, M., and Harwood, D. (1996). A comparative study of texture measures with classification based on featured distributions. *Pattern Recogn.* 29, 51–59.

[38] Ahonen, T., Hadid, A., and Pietikinen, M. (2006). Face description with local binary patterns: application to face recognition. *IEEE Trans. Pattern Anal. Mach. Intell.* 28, 2037–2041.

[39] Freund, Y., and Schapire, R. E. (1995). A decision-theoretic generalization of on-line learning and an application to boosting. *Comput. Learn. Theory* 55, 23–37.

[40] Moreno, R. J. (2014). "Robotic explorer to search people through face detection," in *Proceedings of the CIIMA International Congress of Engineering Mechatronics and Automation* (Rome: IEEE), 1–4.

[41] Abaya, W. F., Basa, J., Sy, and Abad, A. C. (2014). "Low cost smart security camera with night vision capability using Raspberry Pi and OpenCV," in *Proceedings of the HNICEM International Conference on Humanoid, Nanotechnology, Information Technology, Communication and Control Environment and Management* (Rome: IEEE), 1–6.

Biographies

H. T. Nguyen is a Ph.D. candidate in School of Electronics and Computer Engineering, Chonnam National University, Gwangju, Republic of Korea. He received his Bachelor's degree in Electronic & Communication Engineering from Thai Nguyen University of Technology, Vietnam in 2012. His research interests include Computer Vision, Wearable Devices, Robot Vision and Embedded Systems.

S. Y. Na received B.S. degree from Seoul National University in 1977, M.S and Ph.D degrees in 1984 and 1986, respectively, from the University of Iowa, USA. Since 1987, he has been a professor at the department of Electronics and Computer Engineering, Chonnam National University. His current research topics are controller design, soft computation methods, sensor-based control, robotics and pattern recognition.

J. Y. Kim received B.S. degree, M.S degree and Ph.D degree in 1986, 1988 and 1944, respectively, from Seoul National University. From 1993 to 1994 he was engaged as a research engineer at Korea Telecom. Since 1995, he has been a professor at Chonnam National University. His current research topics are audio-visual speech process, image processing and cognitive radio.

G. S. Sun received Ph.D degree in 2015 from Chonnam National University. Since 1996, he has been a professor of Mechatronics Engineering at Korea Polytechnics College, after working at various departments of Kia Motors Company. He published nine books in the field of control applications. Now he is deputy director of Smart Factory work promotion center in Gwangju Metropolitan Office. His current research topics are smart factory, system integration for IoT, and sensor-based factory automation.

Cluster Head Energy Optimization in Wireless Sensor Networks

Shahab Tayeb*, Miresmaeil Mirnabibaboli and Shahram Latifi

Department of Electrical & Computer Engineering, University of Nevada,
Las Vegas, NV, United States
*Corresponding Author: shahab.tayeb@unlv.edu

Received 24 October 2016; Accepted 20 November 2016;
Publication 30 November 2016

Abstract

Wireless Sensor Networks (WSNs) consist of many sensor nodes which are vital to various applications in our daily lives. Optimizing energy usage is a key challenge for WSNs. Improving energy utilization can help with controlling traffic, saving energy and eventually, resulting in a better lifetime. In this paper, a modification of the Energy Efficient Credit-Based routing algorithm is proposed, which selects the most optimal Cluster Head based on the priority of relay nodes. Simulation results demonstrate that the proposed algorithm achieves a more efficient load-balancing, a better lifetime, and lower energy consumption, at the expense of slightly higher packet loss and lower data delivery rate. The results are compared with the commonly used Low-Energy Adaptive Clustering Hierarchy algorithm.

Keywords: Clustering, Energy Efficient Routing, Load-balancing.

1 Introduction

Wireless Sensor Network (WSN) is a distribution of autonomous sensors, which cooperatively monitor physical or environmental conditions, such as temperature, vibration, pressure, sound and so on. WSNs are used in many areas, including home automation, health monitoring or other healthcare

Journal of Software Networking, 137–162.
doi: 10.13052/jsn2445-9739.2016.008

applications, industrial process control and monitoring, etc. Applications of WSNs are expanding and the implementation of multifunctional and reliable WSNs is of utmost importance.

The detection process in WSNs mostly depends on sensor node's physical conditions and the solutions of detection problems are largely based on hardware rather than software. After detection, the node has to find whom and how to transfer the sensed data. After that, the turn passes to data transfer process. This process doesn't take much effort from sensor node due to the small size of desired data.

Low energy consumption is a critical task in WSNs, especially in sensor networks comprised of nodes that are considered lightweight with limited battery power. The most critical process in sensor networks is the routing because of high energy consumption, end-to-end delay, and control of packet overhead. Thus, it is required to have a routing mechanism for reducing energy consumption in sensor nodes and for increasing the network lifetime. The faster is the routing process, the longer is the sensor node lifetime and the less is the energy consumption. Hence, the development of efficient routing algorithms is a crucial task in WSN.

On one hand, low energy consumption is an important limitation in sensor networks, which are comprised of lightweight nodes with limited battery power. Hence, preserving the energy becomes a critical task in such networks. On the other hand, routing is a critical process in sensor networks due to concerns about energy consumption, end-to-end delay, and packet overhead. Thus, it is required to have a good routing mechanism in WSNs for reducing energy consumption in sensor nodes and for increasing the network lifetime. The process of setting up the routes during the initialization is influenced by energy considerations. Furthermore, load-balancing the resources even-handedly prevents bottlenecks from forming and this is another challenging task [1, 2].

To increase the performance of WSN routing, multiple paths can be used concurrently. In coherent routing, the data is propagated after such processing as duplicate prevention, time-stamping, etc. The performance of routing protocols is linked to the architectural model and depends heavily on the implementation model. Design constraints might further impact the performance [3]. The routing protocol, especially with regard to route stability and the minimization of energy consumption, is significantly influenced by its data delivery model (Figure 1).

When it comes to routing in WSNs, energy consumption is not the only consideration. Such parameters as reliability, end-to-end delay, throughput,

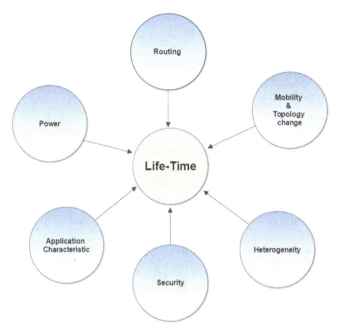

Figure 1   Illustration of parameters affecting routing.

or other QoS metrics are important when selecting the routing. These measures are directly related to bottleneck avoidance or to load-balance the energy consumption for increasing node and/or network performance and lifetime. Avoiding bottlenecks depends on the traffic load, which is in turn influenced by load balancing. Improper load-balancing will lead to unstable routes, which adds to the energy burden on sensors nodes and potentially deteriorates the network. Another consequence of improper load-balancing is partitioning the network into two or more segments. Energy consumption is one of the constraints and some literature [4] claim Dynamic Source Routing (DSR) protocol is more suitable in terms of small energy density.

The remainder of this paper is organized as follows. Section 2 discusses the related works on routing protocols and the issues associated with these algorithms in terms of load-balancing, bottleneck prediction, and congestion avoidance and management. In Section 3, the proposed algorithm is introduced, which incorporates load-balancing, congestion avoidance, and bottleneck avoidance. The performance and evaluation of our proposed algorithm, results and discussions are presented in Section 4. In Section 5, conclusion and future work, as well as recommendations, are presented.

2 Literature Review

The routing protocol performance is a function of to the architectural model, for sensor networks have been considered different architectures and design constraints [4]. For route stability, energy and bandwidth are important optimization factors. There are different setups that utilize mobile sensors and the most part of network architectures sensor nodes are assumed as stationary. Node deployment affects the performance of the routing protocol and it is application dependent. The deployment can be either self-organizing or deterministic. In self-organizing systems, infrastructure is created in an ad hoc manner and the sensor nodes are scattered randomly. In deterministic situations, data is routed through pre-determined paths and the sensors are manually placed. The position of the cluster-head or the sink is also crucial in terms of energy efficiency and performance.

The process of setting up the routes during the creation of an infrastructure is influenced by energy considerations. If all nodes are in close proximity of the sink, direct routing would perform well enough. But in general multi-hop routing becomes unavoidable as sensors are scattered randomly over an area of interest. The routing protocol, especially with regard to route stability and the minimization of energy consumption, is highly influenced by the data delivery model. The sensor node in a sensor network can be associated with various functionalities. A node can be dedicated to a special function such as sensing, aggregation, relaying, depending on the application. As a result, the energy of node might quickly drain in case of the three functionalities engaging at the same time and it is a critical aspect in perspectives of node capability. To reduce the transmission in sensor nodes is using data aggregation when similar packets from multiple nodes are aggregated. For that purpose, sensor nodes are generating significant redundant data. In some sensor network applications, it is only required the success of messages delivery between a destination and a source. Other ones need even more assurance. The requirements of the real-time delivery and the maximization of network lifetime are:

a. Non-real time delivery: The message delivery assurance is indispensable for all routing protocols, which means that the protocol always has to find the route (if it really exists) between the communicating nodes.

b. Real-time delivery: Sometimes it is required that a message is delivered in a specified time [2], otherwise the message's information content is decreasing or the message becomes useless after time bound. Anyway, the main objective of these protocols is the network delay to complete control. These protocols' average-case performance may be evaluated by the time constraints and message delivery ratio measuring.

c. Network lifetime: This is crucial in networks, where the application has to run as long as possible. The metric which is used for determining the network lifetime is application dependent. In most protocols, it is assumed that every node is equally important and in such protocols as a metric is used either the time until the first node dies, or the nodes average energy consumption. But in case of nodes are not equally important, a reasonable metric can be the time till high-priority or the last nodes die.

The prediction process in WSNs mostly depends on sensor node's physical conditions. Prediction solutions are largely hardware-based. After prediction, the routing process takes place; followed by, the node finding the path for transferring the sensor data. After that, the route is selected and data is transferred. Due to the small volume of sensor-generated data, processing doesn't put a burden on the sensor node. Routing paths are established by one of the following methods: reactive, proactive, or hybrid [5, 6]. Proactive protocols compute and determine paths and save them in a table in the memory of each sensor, even before they are needed. Any changes should be propagated throughout the network. Updates to the routing table for each sensor requires large memory and resources because WSNs are comprised of thousands of sensors. In case of reactive protocols, routes are calculated only when they are looked up. Hybrid protocols combine characteristics of both reactive and proactive protocols [5].

Generally, WSN routing protocols are divided into three categories: hierarchical-based, location-based, and flat-based routing. In hierarchical-based protocols, sensors perform different tasks. Scalability is the main sensor attribute in the design of such networks. Hierarchical-based routing operates in two modes. The first mode is used for selecting Cluster Heads (CH) while the second is for routing and identifying a specific event. In location-based routing, sensors' locations are used to route data. Moreover, location and position information are required to compute the distance between two given sensors. This way, the average energy consumption can be determined. Information about the location can be obtained using two techniques: one is to calculate the neighbor node, and the other is using GPS. In flat-based routing, all sensors share the same set of tasks in the network. Sensor selection for querying is difficult due to lack of global ID along with random extensions of nodes [7].

Routing protocols such as Directed-Diffusion (DD) routing [8], Sensor Protocol for Information via Negotiation (SPIN) [9, 10], Low Energy Adaptive Clustering Hierarchy (LEACH) [11], Gradient-based Routing (GBR) [12], Dynamic Source Routing (DSR) [13], and Power-Efficient Gathering in Sensor Information Systems (PEGASIS) [14] were introduced for efficient multi-hop routing in WSNs [15]. SPIN cannot guarantee 100 percent delivery of packets from a source node to the sink node. Moreover, SPIN needs a complete knowledge of the topology. DD is a data-centric protocol and it is mandatory for the sink node to create, transfer, and re-route the intermittent updates packets. The aim of designing DD was efficiency in energy consumption, resulting in an increase in network life expectancy. To reduce energy consumption, DD uses compression and processing of information within the network. However, it has limitation caused by the significant diffusion which results in a high level of overload. Furthermore, each node forms a gradient during the propagation toward all its neighbours. These gradients are paths used to route the packet. However, they provide limited information in the sense that each node is only capable of recognizing its immediate neighbour(s). As a result, DD is susceptible to bottlenecks, which causes inefficiency in energy consumption. Battery-Powered Sensor Nodes (BPSNs) can rarely meet design goals of long network lifetime and high reliability [16].

Directed Diffusion [8] is suitable for most applications but performs weakly for scenarios where there are many receivers and reference points. When receivers are related to one other, the data volume increases significantly. Gradient-based routing is a modified version of DD [9, 17]. This solution uses such techniques as data aggregation and congestion management in order to balance the traffic uniformly, which helps in balancing the load on sensor nodes and thus, increases the network lifetime [17]. The LEACH routing algorithm is characterized as hierarchical and is designed to gather and receive data from and to the sink node, which essentially acts as a base station for ad-hoc networks [18, 19]. PEGASIS and LEACH algorithms are similar. They both use a multi-hop algorithm for routing while a single sensor node is selected for forwarding to the sink node. Each node is a member of a chain for forwarding packets, resulting in a decrease in overhead. Using dynamic clustering in PEGASIS, performance increases three times.

2.1 Classic Protocols for Improving Load-Balancing

Clustering protocol is the commonly used algorithm focusing on lifetime, energy, and load balancing. Clustering offers scalability and location awareness while supporting mobility and data aggregation [19]. Data aggregation is combining data packets from multiple sensors in a single packet, using functions such as min, max, average, or duplicate removal. Data aggregation controls the load-bar which results in a decrease in the total number of packets.

In clustering algorithms, the load is balanced via dynamic selection of CH which provides good balancing of the energy of sensor nodes [20]. By the help of cluster rotation, CH transmits to all sensor nodes, resulting in a balanced consumption of energy throughout sensor lifetime. However, because of using multi-hop transition, CH located near the sink, have to transmit more traffic than other CH which creates such issues as bottleneck and congestion. As a result, the closest CH will be lost sooner than other CH [21]. Congestion avoidance, bottleneck prediction, and load-balancing are the key challenges in the WSN. Optimizing selection of sensor nodes helps with congestion management and facilitates load-balancing. For bottleneck prediction throughout network lifetime, sensors are controlled via monitoring buffer capacity and channel usage. On the other hand, congestion management and avoidance mechanisms can increase the performance of network and balance traffic load in multi-hop routing [18].

2.2 Control Plane

Reference [22] proposed an algorithm where the sensor node is aware of its own geographical location, using GPS, and is also aware of all its neighbours' locations. Additionally, the node is aware of the sink node. An enhanced version of this algorithm is proposed in Ref. [23], which utilizes buffer capacity and compares the node buffers. This facilitates congestion avoidance and prediction.

In Ref. [24], multi-path routing, i.e. load-balancing, is used in order to increase the throughput of the network. Moreover, the proposed algorithm is a hybrid method which performs congestion avoidance using traffic-aware routing. It also normalizes queue length and the depth of sink node from sensor nodes. In other words, it maintains a routing table in order to the least depth, or the shortest route.

Reference [25] proposed a fair cooperative routing method for heterogeneous overlapped WSN. To retain the total energy, this paper introduced an energy pool and used a cooperative packet-forwarding mechanism acting as an agent for fair cooperation. In Ref. [26], an application-driven algorithm was used based on energy-efficient node selection. This algorithm introduced an application-driven development which increases network lifetime and supports load-balancing via grouping sensor nodes running similar applications. Despite the increased delay, an improvement in load-balancing and lifetime is achieved. Reference [27] proposed a hybrid nature-inspired optimizer for WSN where they introduced three multi-objective models for the planning stage, namely, Load-Balanced Model, Interference Model, and Flow-Capacity Model. In Ref. [28], they presented a method based on the results of a two-year deployment. This included 455 wireless energy plug-load devices as well as seven load-balancing routers, which were implemented into a building for one year. Abdelhakim et al. [29] proposed an energy-efficient algorithm for time-sensitive application using optimal topology design. Ren et al. [30] analysed network lifetime for BPSN and proposed an energy-based analytical model. Deva et al. [31] introduced deputy CH for alternate path selection, resulting in higher energy efficiency and throughput for WSNs.

3 The Proposed Algorithm

In Ref. [2], a clustering algorithm called Energy Efficient Credit-Based (EECB) is introduced, a clustering algorithm which uses candidate nodes for transmission between clusters. The relay node provides an optimum route for energy efficiency. The suggested method for an optimized routing selection with division of duties between nodes in the method of Directed-Diffusion, changes the flat routing into the hierarchical one. The change continues to achieve a suitable consumption of energy and ignorance of nodes which are not on the way of sink-source. This prevents a waste of energy. In fact, the transformations of flat routing into hierarchical routing causes a condition for the method of Directed-Diffusion, where we can insert different duties for nodes, based on practical policies, and divide the nodes, based on duties. At the beginning, the duty division takes place based on local clustering [2] and the nodes compete to be a cluster head. After a period, a node with the highest remaining energy than its geographical area nodes is selected as the cluster head node. If the reminder energy of the cluster head node is less than the threshold, it will enterprise a new cluster replacement and the competitor node with maximum reminder energy will be replaced by previous cluster head node. It means that the sensors with maximum reminder energy are selected as a cluster (Figure 2). This is like some sensors get very close locations. Thus, the sensor selects the wireless with the maximum reminder energy as the first cluster. This selection can be done based on Equation (1) of selection

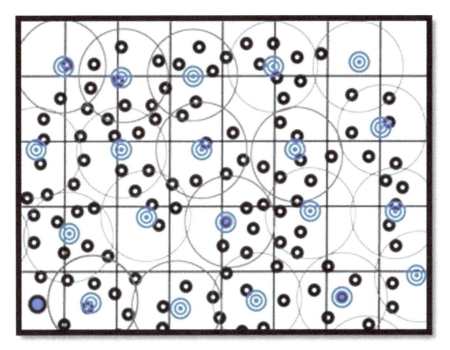

Figure 2 Illustration of arrangement and clustering of network nodes.

of alternative cluster head (Figure 2). Each wireless sensor with a maximum factor other than primary cluster head is selected as the alternative cluster head.

$$CH\_R_F = E_{Ni} - \log_{10}(d_{Ni \text{ to primary } CH})$$
$$- \log_{10}(AVG\_D_{Ni \text{ to primary } CH}) \quad (1)$$

where E_{Ni} is the energy of the candidate node for replacement, d_{Ni} to primary CH is the distance between an alternative node to the primary cluster head. AVG_D shows the delay between the alternative node and the cluster head.

When a node senses an event, a packet named event with a determined lifetime is created and it is sent to the cluster head. The cluster head node sends the discovered packet to the adequate relay node by use of a specified function, in order to send the adequate cluster head. The research will deal with this issue in the next parts of this dissertation; there is an important point in sending packet by cluster. In other words, if the sink node is in the sending limitation of cluster head radius or there is not any relay node, data will be sent to the sink node directly. Otherwise, the cluster head selects the node with the highest amount of route choice as the next hop. Figure 3 demon-strates how clustering takes place and Figure 4 proposes a flowchart for CH replacement.

In this part, we introduce some parameters for selecting the best relay node and after that, making a list of relay nodes based on priority. The priority of relay nodes includes the following parameters.

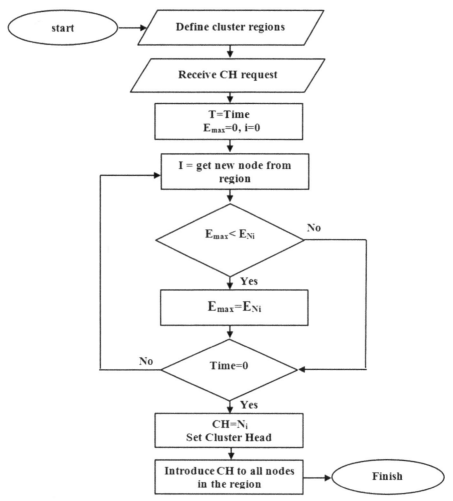

Figure 3   Flowchart of clustering.

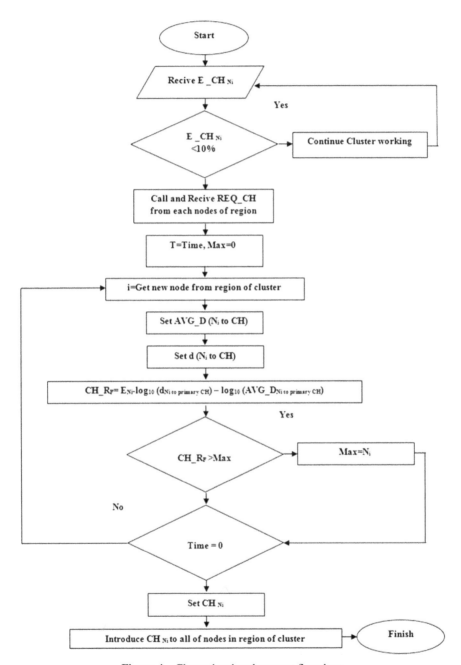

Figure 4 Cluster head replacement flowchart.

3.1 Candidate Relay Nodes (C$_N$)

Cluster Head (CH) selects the best nodes for data transmission based on the estimate the send-radius. In this paper, Equation (2) is used to select the set of candidate nodes. Using Equation (2), the distance between the current node and the CH is estimated. It is then subtracted from the send-radius (x_s). If the result is within the pre-defined threshold, the node is selected as a candidate relay node. The details are illustrated in Figure 5.

$$C_N = \{C_i| \rightarrow |(d_{Node\,.\,to\,.\,CH} - x_s)| \leq \lambda\} \tag{2}$$

where C_i represents a given node.

3.2 Feasibility Condition (FC)

The distance from the candidate node to the sink node: The distance of the node i to the sink is designated as d_{Ni}. It should be noted that each node is aware of its own distances to the sink node. The relationship between the distance and mean energy consumption is reciprocal; i.e. a more distant node has a higher mean of energy consumption which causes more delay for establishing adjacency with the sink node. As it can be seen in Figure 6, node A is closer to the sink node, than node B. If B is selected for transmission, the routing

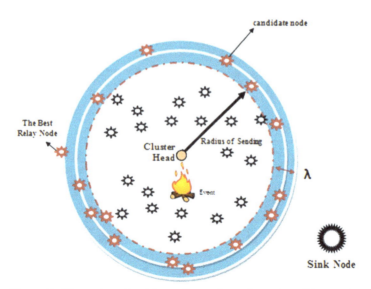

Figure 5   Illustration of node's sending radius and set of candidate nodes.

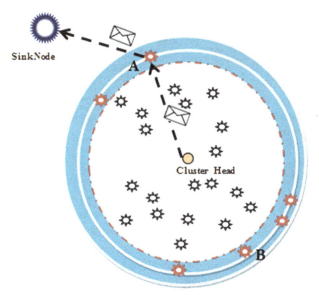

Figure 6 Effect of sending the distance to sink node.

map will change completely and the sink will receive the packet with a higher delay.

The distance of candidate node from the sender node: $d_{Ni.to.CH}$ is the distance of a given node to CH.

The remainder energy in the candidate node: E_R is the remainder of energy.

The amount of available buffer memory in the candidate node: the available space in the buffer memory is calculated using Equation (3).

$$M_F = M_t - M_e \tag{3}$$

where M_F is the available buffer, M_t is the total capacity of buffer and M_e is the existing data in the buffer.

End-to-end delay: End-to-end delay is calculated using Equation (4).

$$D_{end-end} = (D_{trans} + D_{prop} + D_{proc}) \tag{4}$$

where $D_{end-end}$ is the end-to-end delay, D_{trans} is the transmission delay whereas D_{prop} and D_{proc} are the propagation and process delay, respectively.

The ratio of failure/success: The ratio of failure to success in each time period is another parameter to recognize a suitable node. In this article, a self-variable memory for each sensor node is used, which is called S_{li}.

Each successful transmission results in a one-unit increase in the S_{li} amount. The ratio of success or failure can be calculated using Equation (5).

$$S_s = \left(\frac{\sum S_{li}}{\sum N_i} \right) \times 100 \tag{5}$$

Node selection is carried out using Equation (6). Figure 7 illustrates this offline process step-by-step:

$$F_i = E_{Ri} + \alpha_1 \log d_{Ni.to.\sin k} - \alpha_2 \log d_{Ni.to.CH}$$
$$- \alpha_3 \log M_F - \alpha_4 \log D_{end.to.end} - \alpha_5 \log S_s \tag{6}$$

F_{max} is the selected node and the route selection process depends on the choice of a relay node. After selection, K_{ij} is formed where i is the node number and j is the cluster number. This process is repeated for every other cluster. Each relay node is selected from the set of candidate nodes. Sensor R_i detects a destination object O_i if it is one of the candidate relay nodes. Equation (7) shows the binary model of route selection.

$$\text{Route (s)} \leftarrow \begin{cases} 1, & F(R_i, O_i) \leq r \\ 0, & F(R_i, O_i) > r \end{cases} \tag{7}$$

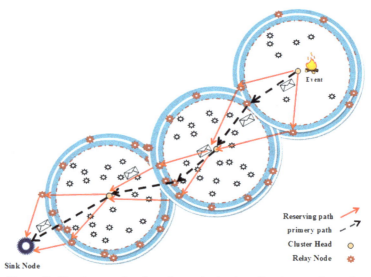

Figure 7　Routing map by clustering and selection of optimum relay node.

where F is the selection function for relay node R_i, the sensed target object O_j and r is the sending radius. When the target object is included, the route function(s) equals to 1, otherwise is 0.

4 Experimental Results and Analysis

To implement the proposed algorithm, MATLAB was used. 200 sensor nodes were stochastically distributed into 100 m². After that, these nodes were divided into several clusters, illustrated by different colours. Then, routing map is established based on the LEACH algorithm (Figure 8). The best relay nodes are selected based on Equation (5). Figure 9 demonstrates the optimum path selection. Figure 10 represents the establishment of routing map using the proposed algorithm and Figure 11 optimizes the path selection, accordingly. In the simulations, five clusters were formed after random distributions of nodes. Three routes were found, which are used alternatively based on the Markov chain (Figures 11–13). Figure 12 performs route selection using Markov chain and Figure 13 chooses the optimum path from Figure 12. The gathered information is sending via wireless sensor network to monitor or controller that is called sink. The Sink can use the data and event locally from its environment, or it can resend data to other nodes or station from wireless network.

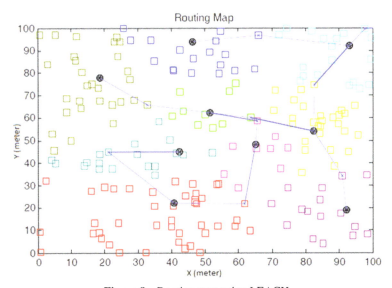

Figure 8   Routing map using LEACH.

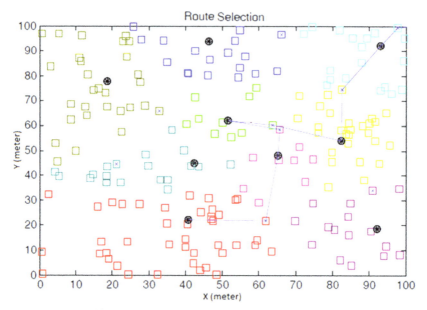

Figure 9 Optimum path selection using LEACH.

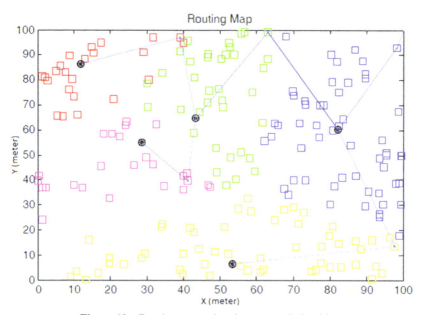

Figure 10 Routing map using the proposed algorithm.

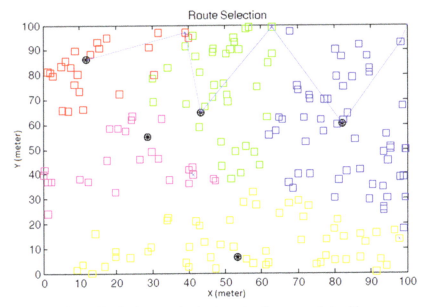

Figure 11 Optimum path selection using the proposed algorithm.

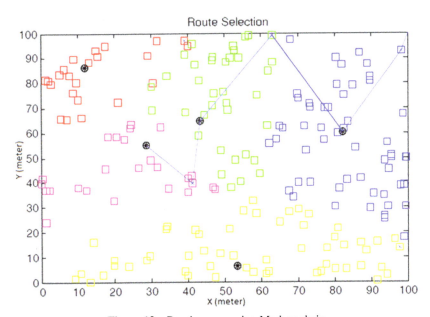

Figure 12 Routing map using Markov chain.

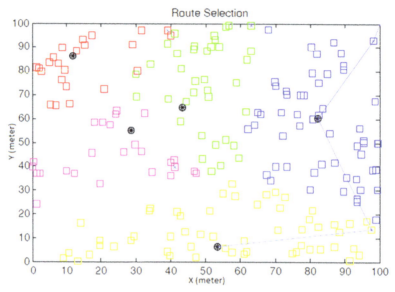

Figure 13 Optimum path selection using Markov chain.

5 Results

For implementations of new routing algorithms for WSNs NS-allionone-2.29 simulator with Diffusion 3.2.0 code have been chosen. NS-2 is used to simulate WSNs and the new algorithms have been implemented in it. About 300 nodes distributed throughout about $400 \times 400 \ m^2$ have been used for these implementations. The 802.11b protocol is used for simulation of wireless scenario Diffusion 3.2.0 in NS-allinone-2.29. The nodes have been randomly expanded in grid according to energy consumption in PCM-CIA WLAN card as NS-2. At the same time some simulations are done by using MATLAB. Simulation parameters are given in Table 1.

Table 1 Simulation parameters

Parameter	Value	Parameter	Value
Layer 3 Protocol	Diffusion	Radio Propagation	Two-way
Layer 2 Protocol	IEEE 802.11g	Packet Size	4 Kbit
$E_{Tx-elec}$	50 nJ/bit	Data Rate	1 Mbps
$E_{Rx-elec}$	50 nJ/bit	Radio Range	90 meter
Sensing Power	4 nJ/bit	Sensing Range	$13 \sim 48$ meter
Area	$160 \times 160 \ m^2$	Number of Nodes	250

Figure 14 compares the remainder energy between LEACH and the proposed algorithm. The energy of LEACH nodes diminished by the 350th time slice whereas the proposed algorithm reduced the energy consumption. Figure 15 illustrates the number of packets lost per time slice, which is lower

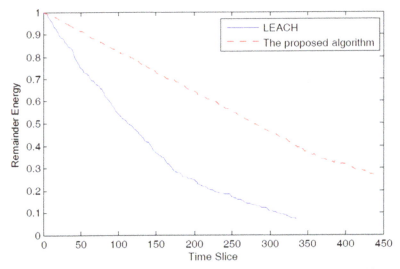

Figure 14   Energy consumption between LEACH and the proposed algorithm.

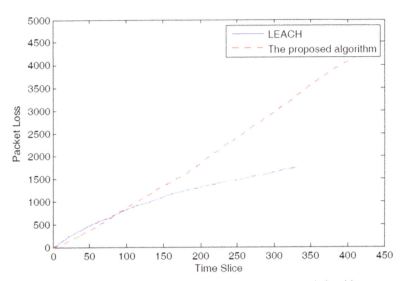

Figure 15   Packet loss between LEACH and the proposed algorithm.

in the proposed algorithm in the first 100 time slices but experienced a steady growth, surpassing the packet loss achieved by LEACH. This is proportionally reflected in the delivery times presented in Figure 16.

Figure 17 summarizes the lifetime obtained using the proposed algorithm, highlighting a lower number of dead nodes in any given time slice as compared to LEACH.

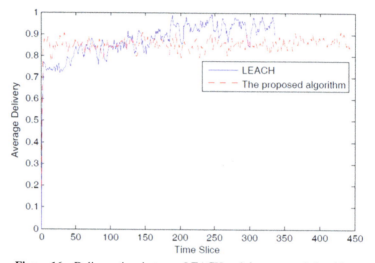

Figure 16   Delivery time between LEACH and the proposed algorithm.

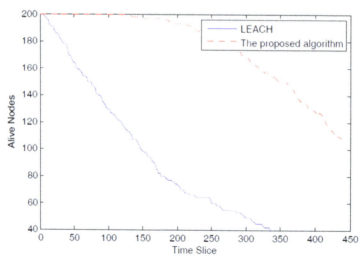

Figure 17   Live nodes between LEACH and the proposed algorithm.

6 Conclusion

In this paper, we proposed a method achieving a more efficient energy utilization by sensor nodes. The simulation results demonstrate that the proposed algorithm could save energy. The method improved the network lifetime and energy consumption, at the expense of packet loss and data delivery. When the network started, this method achieves a higher data-delivery than LEACH algorithm but after the 100th time slice, LEACH performed better in terms of data delivery and packet loss. The proposed method opens up new gateways for future works to optimize energy usage by cluster heads and nodes in WSNs.

Acknowledgment

This work is supported in part by Doctoral Graduate Research Assistantship from UNLV Graduate College and in part by NSF award #EPS-IIA-1301726 (EPSCoR NEXUS).

References

[1] Khan, N. A., Saghar, K., Ahmad, R., and Kiani, A. K. (2016). "Achieving energy efficiency through load balancing: a comparison through formal verification of two WSN routing protocols," in *13th International Bhurban Conference on Applied Sciences, and Technology (IBCAST)* (Rome: IEEE), 350–354. doi: 10.1109/IBCAST.2016.7429901.

[2] Mirnabibaboli, M. (2013). *The Enhancements of Network Lifetime and Supporting Mobility by Utilizing Suitable Routing in the Wireless Sensor Network*. Ph.D. Dissertation, National Academy of Sciences of Armenia, Armenia.

[3] Dugaev, D., Zinov, S., Siemens, E., and Shuvalov, V. (2015). "A survey, and performance evaluation of ad-hoc multi-hop routing protocols for static outdoor networks," in *International Siberian Conference on Control, and Communications (SIBCON)* (Rome: IEEE), 1–11. doi: 10.1109/SIBCON.2015.7147048.

[4] Brar, G. S., Rani, S., Chopra, V., Malhotra, R. Song, H. and Ahmed, S. H. (2016). Energy efficient direction-based PDORP routing protocol for WSN. *IEEE Access*, 4, 3182–3194. doi: 10.1109/ACCESS.2016.2576475.

[5] Jiang, Q., and Manivannan, D. (2004). "Routing protocols for sensor networks," in *Consumer Communications and Networking Conference (CCNC)* (Rome: IEEE), Las Vegas, NV, USA, 93–98. doi: 10.1109/CCNC.2004.1286839

[6] Wang, Y. H., Hsu, C. P., Lin, Y. C., Kuo, C. S., and Ho, H. Y. (2007). "A Routing Method by Reactive Energy Decision in WSNs," in *21st International Conference on Advanced Information Networking, and Applications Workshops, AINAW '07* (Rome: IEEE), 701–770. doi: 10.1109/AINAW.2007.49.

[7] Salehian, S., Masoumiyan, F., and Udzir, N. I. (2012). "Energy-efficient intrusion prediction in WSN," in *International Conference on Cyber Security, Cyber Warfare, and Digital Forensic (CyberSec)* (Rome: IEEE), 207–212.

[8] Intanagonwiwat, C., Govindan, R., and Estrin, D. (2000). "Directed diffusion: a scalable, and robust communication paradigm for sensor networks," in *Proceedings of the 6th Annual International Conference on Mobile Computing, and Networking (MOBICOM '00)* (Rome: IEEE), pp. 56–67.

[9] Jain, V., and Khan, N. A. (2014). "Simulation analysis of directed diffusion and SPIN routing protocol in wireless sensor network," in *2014 Conference on IT in Business, Industry and Government (CSIBIG)*, Indore, India, 2014, pp. 1–6. doi: 10.1109/CSIBIG.2014.7056990.

[10] Heinzelman, W. R., Kulik, J., and Balakrishnan, H. (1999). "Adaptive protocols for information dissemination in wireless sensor networks," in *Proceedings of the 5th Annual ACM/IEEE International Conference on Mobile Computing, and Networking (MobiCom_99)* (Rome: IEEE), Seattle, WA.

[11] Heinzelman, W. R., Chandrakasan, A., and Balakrishnan, H. (2000). "Energy-efficient communication protocol for wireless microsensor networks," in *Proceedings of the 33rd Annual Hawaii International Conference on System Sciences* (Rome: IEEE), 10. doi: 10.1109/HICSS.2000.926982.

[12] Migabo, M. E., Djouani, K., Kurien, A. M., and Olwal, T. O. (2015). A stochastic energy consumption model for wireless sensor networks using GBR techniques. *AFRICON,* 1–5. doi: 10.1109/AFRCON.2015.7331987.

[13] Faruque, J., and Helmy, A. (2003). Gradient-based routing in sensor networks. *ACM SIGMOBILE Mobile Comput. Commun. Rev. Arch.* 7, 50–52.

[14] Kim, D., Ha, S., and Choi, Y. (1998). "K-hop cluster-based dynamic source routing in wireless ad-hoc packet radio network," in *48th IEEE Vehicular Technology Conference, VTC 98*, Vol. 1, Ottawa, ON, 224–228. doi: 10.1109/VETEC.1998.686541.

[15] Lindsey, S., and Raghavendra, C. S. (2002). "PEGASIS: Power Efficient Gathering in Sensor Information Systems," in *Proceedings of IEEE Aerospace Conference* (Rome: IEEE).

[16] Zonouz, A. E., Xing, L., Vokkarane, V. M., and Sun, Y. (2016). Hybrid wireless sensor networks: a reliability, cost and energy-aware approach. *IET Wirel. Sensor Syst.* 6, 42–48. doi: 10.1049/iet-wss.2014.0131.

[17] Mishra, A. K., R. Us. Rahman, Bharadwaj, R., and Sharma, R. (2015). "An enhancement of PEGASIS protocol with improved network lifetime for WSNs," in *IEEE Power, Communication, and Information Technology Conference (PCITC)*, 142–147.

[18] Bushnag, A., Alessa, A., Li, M., and Elleithy, K. (2015). "Directed diffusion based on weighted Grover's quantum algorithm (DWGQ)," in *Systems, Applications, and Technology Conference (LISAT)*, IEEE Long Island, 1–5.

[19] Chughtai, O., N. Badruddin and Awang, A. (2014). "A congestion-aware and energy efficient traffic Load balancing Scheme for routing in WSNs," in *TENCON 2014 – 2014 IEEE Region 10 Conference*, Bangkok, 1–6. doi: 10.1109/TENCON.2014.7022431.

[20] Abbasi, A., and Younis, M. (2007). A survey on clustering algorithms for WSNs. *Comput. Commun.* 2826–2841.

[21] Gupta, G., and Younis, M. (2003). "Performance evaluation of load-balanced clustering of wireless sensor networks," in *10th International Conference on Telecommunications ICT 2003* (Rome: IEEE), Vol. 2, 1577–1583. doi: 10.1109/ICTEL.2003.1191669.

[22] Du, X., Xiao, Y., and Dai, F. (2006). Increasing network lifetime by balancing node energy consumption in heterogeneous sensor networks. *Wirel. Commun. Mobile Comput.* 125–136.

[23] AlAmri, H., Abolhasan, M., Franklin, D. R., and Lipman, J. (2014). "Optimised relay selection for route discovery in reactive routing." *Ad Hoc Netw.*, pp. 70–88.

[24] Chughtai, O., Badruddin, N., Awang, A., and Rehan, M. (2016). Congestion-aware and traffic load balancing scheme for routing in WSNs. *Telecommun. Syst.*, 1–24.

[25] Ren, F., He, T. Das S. K., and Lin, C. (2011). Traffic-Aware Dynamic Routing to Alleviate Congestion in Wireless Sensor Networks. *IEEE Trans. Parallel Distribut. Syst.* 22, 1585–1599. doi: 10.1109/TPDS.2011.24.

[26] Kinoshita, K., Inoue, N., Tanigawa, Y., and Tode, H. T. (2016). "Watanabe. Fair routing for overlapped cooperative heterogeneous WSNs." *IEEE Sensors J.*, 3981–3988.

[27] Marques, B., and Ricardo, M. (2016). Energy-efficient node selection in application-driven WSN. *Wirel. Netw.*, 1–30.

[28] Benyamina, D., Hafid, A., Hallam, N., and Gendreau, M. (2012). A hybrid nature-inspired optimizer for wireless mesh networks design. *Comput. Commun.* 35, 1231–1246.

[29] Abdelhakim, M., Liang, Y., and Li, T. (2016). Mobile coordinated wireless sensor network: an energy efficient scheme for real-time transmissions. *IEEE J. Select. Areas Commun.* 34, 1663–1675. doi: 10.1109/JSAC.2016.2545383.

[30] Ren, J. Zhang, Y., Zhang, K., Liu, A., Chen J., and Shen, X. S. (2016). Lifetime and energy hole evolution analysis in data-gathering wireless sensor networks. *IEEE Trans. Ind. Inform.* 12, 788–800. doi: 10.1109/TII.2015.2411231.

[31] Deva Sarma, H. K., Mall, R., and Kar, A. (2016). E2R2: Energy-efficient and reliable routing for mobile wireless sensor networks. *IEEE Syst. J.* 10, 604–616. doi: 10.1109/JSYST.2015.2410592.

Biographies

S. Tayeb received the M.S. degree (Magna Cum Laude) in radio engineering and communications and the B.S. degree (Magna Cum Laude) in telecommunications engineering from the State Engineering University of Armenia in 2012 and 2010, respectively. He is currently a Ph.D. candidate in the department of electrical and computer engineering at University of Nevada Las Vegas (UNLV). He holds CCIE R&S, CCDP, CCNP R&S, and CCAI from Cisco; CNSS 4011 Recognition from NSA; TKT from Cambridge; and VMCA-DCV from VMware. Prior to joining UNLV, he was an instructor and instructor trainer at Cisco Networking Academy where he was recognized as top %5 expert level instructors globally. He has been invited to deliver instructor-level courses in various countries around Europe, Middle East, Africa, and North America. He has authored/co-authored several research papers on network security, Wireless Sensors Networks, Internet of Things, and Big Data. His research interests span the areas of Internet of Things, Information Assurance, Security, and Wireless Sensor Networks utilizing such tools as Deep Learning and Big Data Analytics. He is a member of IEEE, ISOC, Teachers without Borders, and NSPE.

M. Mirnabibaboli was born in Babol, Iran, in 1984. He received the B.S. degree in Software Engineering from the University of Sari in 2007. His fields in M.S. and Ph.D. were Information Technology-Network Engineering from IT department of QIUZ and NASRA. He taught Computer courses for B.S. and M.S. students at universities of IAU, PNU, UMZ, and UAST in Iran between 2009 and 2015. He was adviser and member of Informatics Councils of IT in AgriBank of Iran in 2014 and 2015.

In 2016, he joined the Department of Electrical and Computer Engineering at University of Nevada, Las Vegas, as a PostDoc-Research Scholar. He researched on WSNs, Data Mining, Bio-Data Mining and Cloud Computing in the past 8 years.

S. Latifi, an IEEE Fellow, received the Master of Science degree in Electrical Engineering from Fanni, Teheran University, Iran in 1980. He received the Master of Science and the Ph.D. degrees both in Electrical and Computer Engineering from Louisiana State University, Baton Rouge, in 1986 and 1989, respectively. He is currently a Professor of Electrical Engineering at the University of Nevada, Las Vegas. Dr. Latifi is the co-director of the Center for Information Technology and Algorithms (CITA) at UNLV. He has designed and taught undergraduate and graduate courses in the broad spectrum of Computer Science and Engineering in the past three decades. He has given seminars on cyber-related topics all over the world. He has authored over 250 technical articles in the areas of networking, cybersecurity, image processing, biosurveillance, biometrics, document analysis, fault tolerant computing, parallel processing, and data compression. His research has been funded by NSF, NASA, DOE, DoD, Boeing, Lockheed and Cray Inc. Dr. Latifi was an Associate Editor of the IEEE Transactions on Computers (1999–2006), an IEEE Distinguished Speaker (1997–2000), and Co-founder and General Chair of the IEEE Int'l Conf. on Information Technology (2004–2015). Dr. Latifi is the recipient of several research awards, the most recent being the Silver State Research Award (2014). He is also a Registered Professional Engineer in the State of Nevada.

Implementing a Stereo Image Correction Method for 3D Surgery Microscopes

Jiyoung Yoon[1], Nakhoon Baek[2,*], Cheolhwan Kim[3]
and Hwanyong Lee[4]

[1]*Institute of Advanced Convergence Technology, Kyungpook National University, South Korea*
[2]*School of Computer Science and Engineering, Kyungpook National University, South Korea*
[3]*School of Electronic Engineering, Kyungpook National University, South Korea*
[4]*Ajou University, South Korea*
E-mail: {wngus235; oceancru; hwanyong.lee}@gmail.com; kch1@knu.ac.kr
*Corresponding Author

Received 10 November 2016; Accepted 24 November 2016;
Publication 6 December 2016

Abstract

Recently 3D scenes are used in various industrial fields such as medical applications, computer games, surface examinations, biology and others. 3D optical microscopes can show extremely precise details in 3D. To reconstruct the 3D images of an optical microscope, two cameras are mounted on the optical microscope. Incoming images through an object lens of an optical microscope are projected on sensors of mounted cameras by using refraction mirrors. Two cameras capture the left and right images to reconstruct the final 3D images. In this paper, we correct the 3D reconstruction errors with the SURF algorithm. We also design a hardware system to correct wrong mirror positions using servo motors. In addition, we propose the surgical system with HMD and wireless communications. This system would be helpful for doctors due to make doctors comfortable and it can be used to educate surgery procedures.

Keywords: Stereo Image, Optical Microscope, Computer Vison, SURF, Medical Device.

Journal of Software Networking, 163–176.
doi: 10.13052/jsn2445-9739.2016.009

1 Introduction

Recently 3D scenes are used in various industry fields. 3D optical microscope [1] is being developed to see much precise details in 3D. The 3D microscopy is widely used in industrial, medical and biological studies. In the industrial field, this system is used to identify errors of the electronic circuits or detects the very small mechanical components. In the medical areas, it is used for actual medical procedures such as polyp removal. It can be used in hair transplant, ophthalmology and biology to identify such cells, chromosomes in research. A microscope with augmented reality technology is being researched to provide a guide to the surgical region and to show the necessary information [10]. Figure 1 is various use cases of imaging devices in medical procedures [2, 3].

Medical doctors make surgery using input images through cameras. They use an optical microscope or an endoscope equipped with cameras. These systems display the expanded images on a monitor for surgical sites it required. To show 3D images of an optical microscope, two cameras are mounted on the optical microscope. Incoming images through an object lens of an optical microscope are projected on sensors of mounted cameras using refractive mirrors. Those refractive mirrors are used to avoid obscuring a view of ocular lens. Two cameras create the left and right images and reconstruct the 3D images. At this time, it is not an easy task, because of the optical design errors and the deformation mechanism of optical microscopes.

In this paper, we corrected those error with the SURF algorithm and designed the hardware system to correct wrong mirror positions using

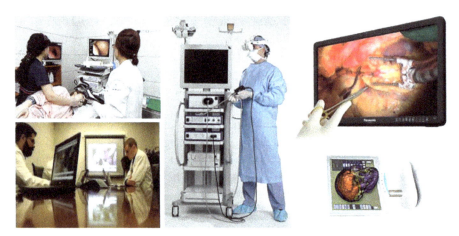

Figure 1   Various use cases of imaging devices in medical procedures.

servo motors. In addition, we propose the surgical system with HMD and wireless communication. Doctors can check surgical position and external circumstance by mounting a camera on the front of the HMD. By sending 3D microscope images through wireless communications, doctors are comfortable with surgery and this system can be used for surgical training.

2 Image Output Error of the Optical Microscope

To show the 3D images from the optical microscopes, two cameras are mounted on an optical microscope. Incoming images through an object lens of an optical microscope are projected on sensors of mounted cameras by using refractive mirrors. Figure 2 shows an example of 3D microscope components.

To send the picture to two cameras on the optical microscope, incoming image into the ocular lens is projected on the camera sensors [4] such as CCD or CMOS. The camera sensor sizes depend on the type of an optical microscope lens and we use about 1/3" to 2/3" sensors and the image of an optical microscope is projected by the camera sensor to adjust the focal length. We can look 3D images of 3D optical microscope on various display

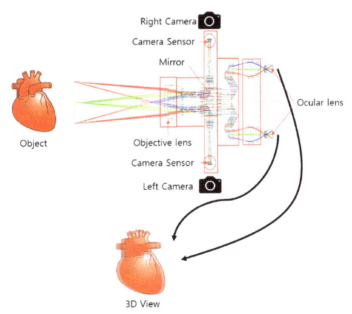

Figure 2 An example of 3D microscope components.

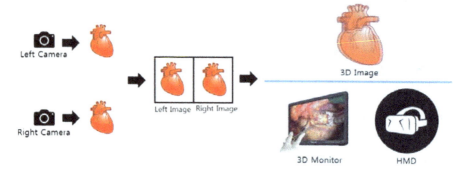

Figure 3 Example of displaying devices of 3D optical microscope.

devices such as 3D monitors, embedded displays, Head Mounted Displays (HMD) [5] and etc. Figure 3 is an example of displaying methods of 3D optical microscope.

We developed the 3D optical microscope. We mounted two 1280×720 resolution cameras on the microscope. We used an HMD system with two 1280×720 resolution microdisplay modules for testing. However, it had the image output error. The image output error of our optical microscope occurred in the cause of the refractive image on the mirrors [6]. The error was generated the image is not projected on the centre point of the camera's sensor because of refraction error of mirrors. It made transformed images. This was caused by the optical design errors and deformation mechanism of the optical microscope. This error can occur during the development of 3D optical microscope projecting images using the mirror. If this error occurs in the optical microscope image output, it is difference the real eye view image through ocular lens and the camera image. Then, the user cannot see the desired image and the three-dimensional image. Figure 4 shows the example of stereo image error on optical microscope.

Stereo image error makes that users can have dizziness and nausea. Then users cannot treat patients and the system is not available. So stereo image error should be corrected for 3D optical microscope system.

3 Stereo Image Correction for 3D Optical Microscope

To correct incoming error image through the optical microscope, we used a SURF (Speeded Up Robust Features) algorithm [6, 7, 11]. SURF is a robust local feature detector that can be used in computer vision tasks like object

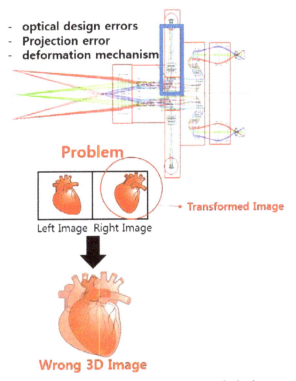

Figure 4   Example of stereo image error on optical microscope.

recognition or 3D reconstruction. SURF is a detector and a high-performance descriptor points of interest in an image where the image is transformed into coordinates, using a technique called multi-resolution. We used SURF algorithm to extract key points and descriptors on left and right image. We could detect the almost same key points and descriptors on the left and right image even if the images were transformed. Then we got a homography [8] by key points and descriptors. We used a homography to correct stereo image error.

To correct stereo images for 3D optical microscope processed in the following order. First, shot the checker image by left and right camera. Second, extract key points and descriptors on left and right images using SURF algorithm. Third, transform left and right images to align centre and perpendicular using key points. Fourth, clip missed image area by transforming images. Finally, fit stereo images at a screen. Figure 5 shows to correct stereo images using checker.

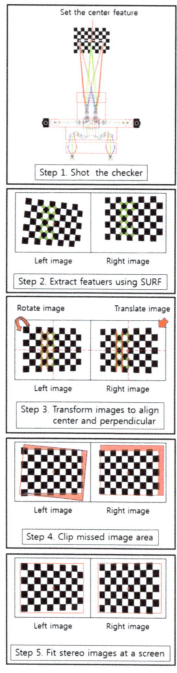

Figure 5 To correct stereo images using checker.

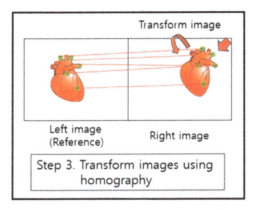

Figure 6   To correct stereo images without checker.

If there is not the checker image, then it should be used other method [6]. We changed Third step in the method described above. We choose the reference image that is left or right image. Then, transform the other image using a homography by feature matching. The fourth and fifth operations are performed in the same manner. Figure 6 shows to correct stereo images without checker.

Using SURF algorithm, it takes about 300~350ms to obtain a homography and clipping region on i7-5930K CPU. Image correction is executed first once using first left and right image frame. Because the deformation mechanism of the camera connector and mirror part on the optical microscope does not occur frequently. This method made a good stereo images. However, this method had two drawbacks. First, it occurs the loss of images on the clipping step. The loss of image due to clipping depends on a transformation of image using homography. But this is very small area and we could get a sufficient image region. Second, it processed to transform left and right images every frame. But it was not big problem. To solve these drawbacks, we should fix the deformation mechanism on the optical microscope.

4 Experiment and Result

We used the devices that were our optical microscope, two Full HD cameras with coaxial cable, the capture board and the personal computer for display and correcting image. The personal computer system was Window 7 64bit and the development tools is Visual Studio 10 MFC. It is not necessary the high performance PC, because image correction is performed once the first time.

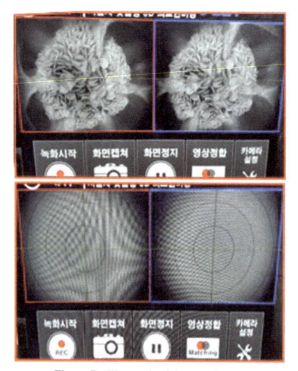

Figure 7   The result of the experiment.

However, it is required the PC with enough performance to display the image in real time. Figure 7 is the result of the experiment.

5 Design Hardware System to Correct Stereo Image Error for 3D Optical Microscope

We designed the hardware system to correct stereo image error for 3D optical microscope because we didn't want to loss of image region and to process transforming images every frame. This system attach servo motors to mirror tilting part of 3D optical microscope. Servo motors move and tilt the mirror to make correct stereo images. Servo motors are controlled by PC system that runs SURF algorithm. The control values of servo motors are reference a homography. We get the transform values and control servo motors by matching stereo image. This system will make correct stereo images without loss of image region. Figure 8 shows hardware system to correct stereo image error for 3D optical microscope.

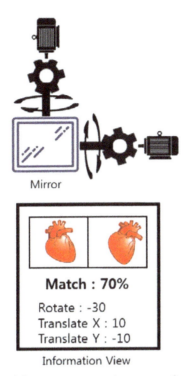

Mirror

Information View

Figure 8   Hardware module to correct stereo image error for 3D optical microscope.

6 Surgery System with HMD and Wireless Communication

We propose the surgical system with HMD and wireless communication. Generally, doctors operated using microscope to see magnified images on a monitor. In this case, doctor's eyes cannot look at the surgical site and it can be unnatural. When doctors watch the monitor installed on the top, he can fell the fatigue of his neck. To solve this problem, a HMD is used instead of a monitor. By using HMD, doctors can freely move and focus on surgery. Figure 9 is the proposed surgery system with HMD.

If the HMD is wired to the system, this cable may interfere with doctor's operation. Therefore, it is necessary to transmit images through wireless communication. When using a general HDM, doctor's views are completely obscured. This problem is solved by using a see-through type HMD and mount a camera on the front of the HMD. By using this, doctors can check the surgical position and the external circumstance. Figure 10 shows wireless HMD system and HMD with front camera.

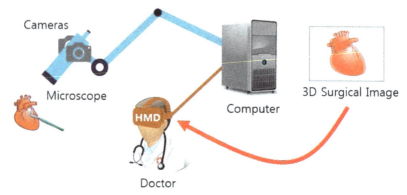

Figure 9   Surgery system with HMD.

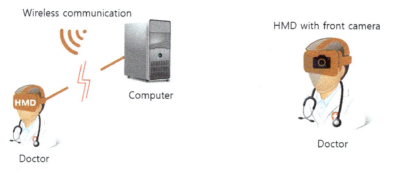

Figure 10   Wireless HMD system and HMD with front camera.

By sending 3D microscope image through wireless communication, doctors are comfortable with surgery. However, if 3D images are transmitted through wireless communication then image output delay may occur. In case of WirelessHD standard, high resolution image could be transmitted at a high speed and at a delay rate of less than 5ms using a broadband of 60 GHz [9]. However, when the wireless communication is interfered, the image is delayed. This can occur a fatal problem for doctors' surgery. Therefore, it is necessary to solve image output delay. When video output delay is a problem in a wireless communication environment, it can be used as a training system for a resident and a student. Residents and students can directly see the doctor's surgical procedures and doctor's scenes outside the operating room. Figure 11 is the proposed surgery and education system with HMD and wireless communication.

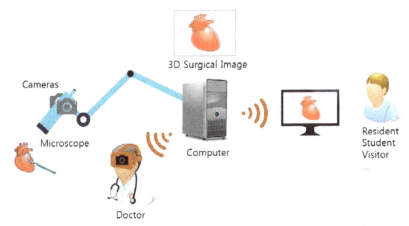

Figure 11 Surgery and education system with HMD and wireless communication.

7 Conclusion

Medical doctors used optical microscopes that shows just a single image. Users do not mind small transformed image errors because this image is used as secondary role to look at things through the ocular lens. However, if there are errors in those 3D images, then users could have dizziness and nausea. If doctors try to treat using 3D display system with stereo image error, they cannot treat patients and the system is not available. So stereo image error should be corrected for 3D optical microscope system. We implemented the stereo image correction system and designed the hardware system for 3D optical microscope. We used the SURF algorithm and got a homography to correct stereo image error. This method made a good stereo images. However, this method had two drawbacks that are decreasing resolution and transforming images every frame. But it was not a big problem and enough to use the system properly. To solve these drawbacks, we designed the hardware system to correct stereo image error for 3D optical microscope. This system will solve stereo image error completely.

In addition, we propose the surgical system with HMD and wireless communication. Doctors can check surgical position and external circumstance by mounting a camera on the front of the HMD. By sending 3D microscope image through wireless communication, doctors are comfortable with surgery and this system can be used for surgical training.

In the future, we will develop the embedded device for this system. We will develop the hardware system that we designed. Additionally, we will make fast embedded device that use SURF algorithm on FPGA to correct stereo image correction.

Acknowledgment

This research was financially supported by the Ministry of Trade, Industry & Energy (MOTIE), Korea Institute for Advancement of Technology (KIAT) and Gangwon Institute for Regional Program Evaluation (GWIRPE) through the Economic and Regional Cooperation Industry.

References

[1] Wikipedia contributors (2015a). *Optical Microscope*. Available at: https://en.wikipedia.org/wiki/Optical_microscope

[2] Figl, M., Ede, C., et al. (2005). "A fully automated calibration method for an optical seethrough head mounted operating microscope with variable zoom and focus," in *Transaction on Medical Imaging* (Rome: IEEE), 1492–1499.

[3] Tamiya, T., Kawanish, M., and Guo, S. (2011). "Skull base surgery using navigation microscope integration system," in *Complex Medical Engineering*, eds J. L. Wu, K. Ito, S. Tobimatsu, T. Nishida, and H. Fukuyama (Berlin: Springer), 185–187.

[4] Wikipedia contributors (2015b). *Image Sensor.* https://en.wikipedia.org/wiki/Image_sensor

[5] Wikipedia contributors (2016). *Head-mounted Display.* https://en.wikipedia.org/wiki/Head-mounted_display

[6] Yoon, J., Kim, H., and Kim, C. (2015). "Development of image error correction system for 3D optical microscope", *Computing Technology and Information Management* (Rome: IEEE), 105–108.

[7] Bay, H., Tuytelaars, T., and Van Gool, L. (2006). Speed-Up Robust Features(SURF), in *Computer Vision – ECCV 2006 Lecture Notes in Computer Science*, Vol. 3951, eds A. Leonardis, H. Bischof, and A. Pinz (Berlin: Springer), 404–417.

[8] Wikipedia contributors (2016). *"Homography_(computer_vision)", Wikipedia, The Free Encyclopedia Wikipedia*. Web.

[9] WirelessHD (2010). *"WirelessHD Specification Version 1.1"*. Available at: www.wirelesshd.org

[10] King, A. P., Edwards, P. J., and Maurer, C. R., et al. (2000). "Stereo augmented reality in the surgical microscope". *Presence* 9, 360–368.

[11] Rublee, E., Rabaud, V., and Konolige, K., et al. (2011). "An efficient alternative to SIFT or SURF," in *Proceedings of the Computer Vision (ICCV), 2011 IEEE International Conference*, Rome.

Biographies

J. Yoon received the M.S. and Ph.D. degree in Computer Science from Kyungpook National University, Korea in 2008 and 2016. He is currently a senior researcher in IACT (Institute of Advanced Convergence Technology), Korea. His research interests include computer graphics standard and parallel processing.

N. Baek is currently a professor in the School of Computer Science and Engineering at Kyungpook National University, Korea. He received his B.A., M.S., and Ph.D. degrees in Computer Science from Korea Advanced Institute of Science and Technology (KAIST) in 1990, 1992, and 1997, respectively. His research interests include graphics standards, graphics algorithms and real-time rendering. He is now also the Chief Engineer of Dassomey.com Inc., Korea.

C. Kim is currently a Ph.D. candidate in the School of Electronic Engineering at Kyungpook National University. His research interests include Computer Vision, 3D Reconstruction and Medical Image processing. He is working in Institute of Advanced Convergence Technology at Kyungpook National University, Korea.

H. Lee received B.S. degree in Computer Science at KAIST and M.S. degree in Computer Science at POSTECH, Ph.D. degree in Computer Engineering at KNU. He is currently an industry cooperation professor at Ajou University. His research interests include computer graphics, image processing, virtual reality and augmented reality.

www.ingramcontent.com/pod-product-compliance
Lightning Source LLC
LaVergne TN
LVHW012330060326
832902LV00011B/1805